MATHEMATICS APPRECIATION

by theoni pappas

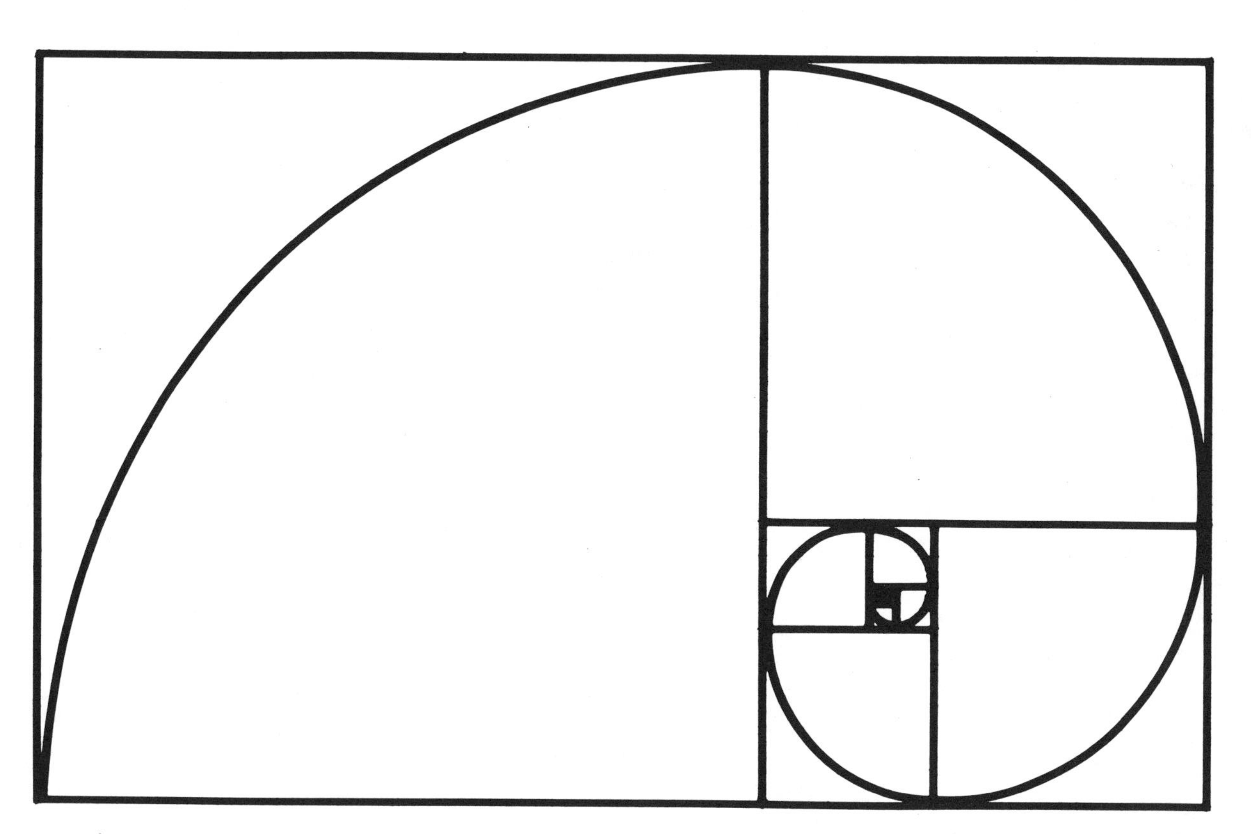

the golden rectangle & the equiangular spiral

TEN COMPLETE ENRICHMENT LESSONS

Revised Edition 1987

November 1992 Printing.

Math Aids/Math Products Plus
P.O. Box 64
San Carlos, CA 94070

Wide World Publishing/Tetra
P.O. Box 476
San Carlos, CA 94070

TABLE OF CONTENTS

INTRODUCTION

In the process of studying or teaching mathematics, we often are so intent on the particular subject matter of a specific course, we neglect to take time to experience the nature of mathematics, its relation to other disciplines, and its beauty. Mathematics has so many facets which touch the different realms of the universe, of the earth, and of our lives.

This book seeks to introduce you to some of the special and different areas in which mathematics appears and has influence. I hope it will help enhance your appreciation of mathematics.

—Theoni Pappas

THE FIBONACCI SEQUENCE

THE FIBONACCI SEQUENCE

BACKGROUND MATERIAL

Fibonacci[1], one of the leading mathematicians of the Middle Ages, made contributions to arithmetic, algebra and geometry. He was born Leonardo da Pisa (1175-1250), son of an Italian customs official, at Burgia in South Africa. His father's work involved travel to various Eastern and Arabic cities, and it was here that Fibonacci became familiarized with the Hindu-Arabic decimal system, which had place value and used the symbol zero. At this time Roman numerals were still being used for calculating in Italy. Fibonacci saw the value and beauty of the Hindu-Arabic numerals, and was a strong advocate of its use. In 1202 he wrote **Liber Abaci**, a comprehensive handbook explaining how to use the Hindu-Arabic numerals; how addition, subtraction, multiplication and division were performed with these numerals; how to solve problems; and further discussion of algebra and geometry. Italian merchants were reluctant to change their old ways; but through their continual contact with Arabs and the works of Fibonacci and other mathematicians, the Hindu-Arabic system was introduced and slowly accepted in Europe.

It seems ironic that Fibonacci is famous today because of a sequence of numbers that resulted from one obscure problem in his book, **Liber Abaci**. At the time he wrote the problem it was considered merely a mental exercise. Then, in the 19th century, when the French mathematician Edouard Lucas was editing a four volume work on recreational mathematics, he attached Fibonacci's name to the sequence that was the solution to the problem from **Liber Abaci**. The problem from **Liber Abaci** that generated the Fibonacci sequence is:

1) Suppose a one month old pair of rabbits (male and female) are too young to reproduce, but are mature enough to reproduce when they are two months old. Also assume that every month, starting from the second month, they produce a new pair of rabbits (male & female).

2) If each pair of rabbits reproduces in the same way as the above, how many pairs of rabbits will there be at the beginning of each month?

⊗=pair, mature enough to reproduce

●=pair, too young to reproduce

$F_n = F_{n-1} + F_{n-2}$

number of pairs

$1=F_1$=1st Fib. no.

$1=F_2$=2nd Fib. no.

$2=F_3$=3rd Fib. no

$3=F_4$=4th Fib. no.

$5=F_5$=5th Fib. no.

[1]Fibonacci literally means son of Bonacci.

Fibonacci did not study this resulting sequence at the time, and it was not given any real significance until the 19th century when mathematicians became intrigued with the sequence, its properties, and all the areas in which it appears.

Fibonacci sequence appears in:

I. The Pascal triangle
II. the golden ratio and the golden rectangle
III. nature and plants
IV. intriguing mathematical tricks

I. The Pascal Triangle

Knowledge of the Pascal triangle goes back prior to the time of Pascal. Pascal wrote a treatise on it in 1653, but it had appeared in various places before this, such as in ***Precious Mirror of the Four Elements*** by Chu Shih Chien in 1303.

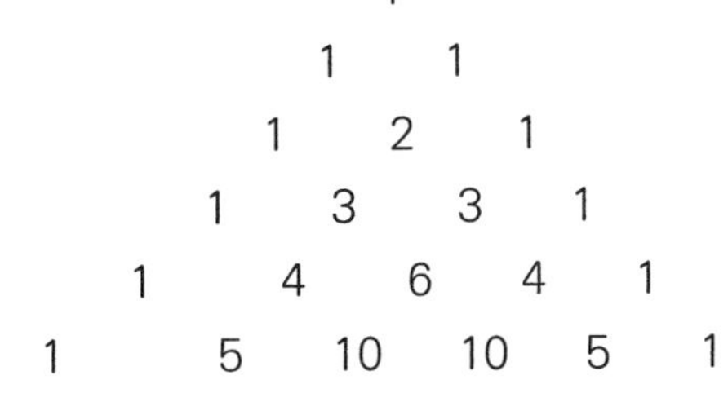

The Pascal triangle is formed as follows: **each number is the sum of the two numbers flanking it in the row above it.**

fascinating properties of the Pascal triangle

1) The other type of numbers found in the Pascal triangle are:

natural numbers-1,2,3,4,5,...
triangular numbers-1,3,6,10,15,21,...
tetrahedral numbers-1,4,10,21,35,...
four-space tetrahedral numbers-1,5,15,35,70,...
five space tetrahedral numbers-1,6,21,56,126,...

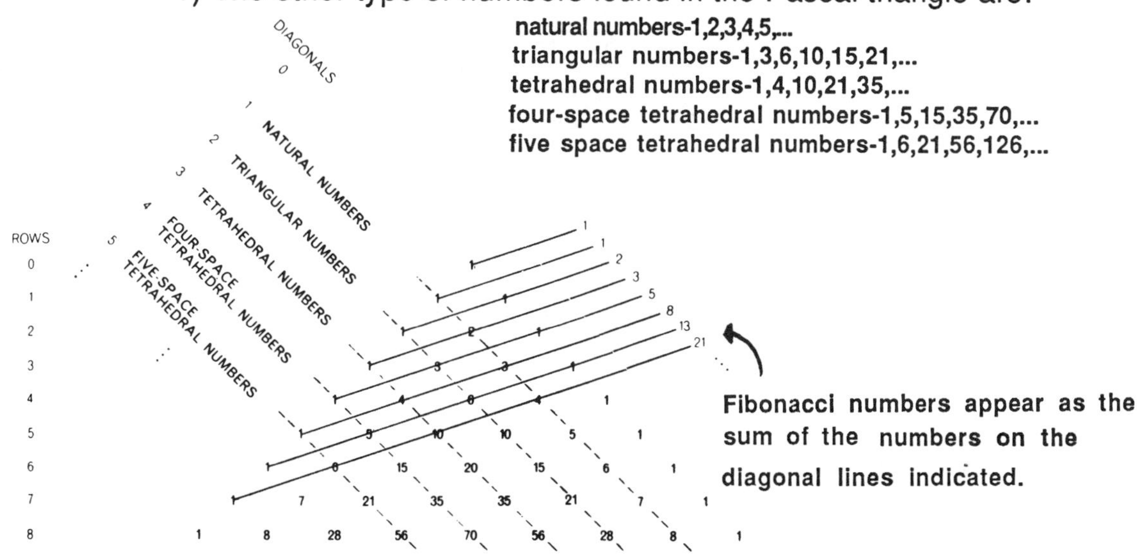

Fibonacci numbers appear as the sum of the numbers on the diagonal lines indicated.

2) The sum of the series of numbers along any negative sloping diagonal in the Pascal triangle is the number below it and to the left, e.g. the sum of the first six triangular numbers is 56.

3) Verner E. Hoggart Jr., professor of mathematics at San Jose State University and editor of **The Fibonacci Monthly,** discovered that by removing the diagonal from the left side of the triangle, partial sums for the Fibonacci sequence are obtained. For example, if the zero diagonal on the left side is removed, then the remaining numbers of the Fibonacci diagonals have sums that are partial sums of the Fibonacci series

$$1=1;1+1=2;1+1+2=4;1+1+2+3=7;1+1+2+3+5=12=5+6+1$$

If the diagonal 0 and 1 are removed from the left side of the Pascal triangle, then the Fibonacci diagonals give partial sums for the series:

$$1+2+4+7+11+17+24+...$$

Some interesting Fibonacci identities[2] are:

1) summation of Fibonacci numbers: $F_1+F_2+F_3+...+F_n = F_{n+2}-1$, where $n>=1$

2) summation of the squares of Fibonacci numbers:

$$F_1^2+F_2^2+F_3^2+...+F_n^2 = (F_n)(F_{n+1}), \text{ where } n>=1$$

3) summation of the odd Fibonacci terms:

$$F_1+F_3+F_5+...+F_{2n-1} = F_{2n}, \text{ where } n>=1$$

4) summation of the even Fibonacci numbers:

$$F_2+F_4+F_6+...+F_{2n} = F_{2n+1} -1, \text{ where } n>=1$$

5) $F_n^2 = (F_{n-1})(F_{n+1})-1$

6) $F_n^2+F_{n+1}^2 = F_{2n+1}$

[2] The proofs of these identities are interesting exercises in proof by induction

7) The Pascal triangle also generates the **binomial formula.** Each row contains the coefficients in the expansion

				1					$=(a+b)^0$
			1		1				$=(a+b)^1$
		1		2		1			$=(a+b)^2$
	1		3		3		1		$=(a+b)^3$
1		4		6		4		1	$=(a+b)^4$
. . . .									$=(a+b)^n =$

binomial formula

$$(a+b)^n = \binom{n}{0}a^n + \binom{n}{1}a^{n-1}b + \binom{n}{2}a^{n-2}b^2 + \ldots + \binom{n}{n}b^n$$

of the binomial, $(a+b)^n$. For example, to find the coefficients for $(a+b)^2$, one looks at the second row from the top row (which is called the zero row, i.e. $(a+b)^0 = 1$) of the Pascal triangle and finds the numbers 1 2 1. These numbers are the resulting coefficients when $(a+b)^2$ is expanded– $(a+b)^2 = \mathbf{1}a^2 + \mathbf{2}ab + \mathbf{1}b^2$.

Since the Pascal triangle is tied into the binomial formula,it is also linked to **probability.** For example, to determine how many combinations there are of 5 different objects taken 3 at a time, look at the 5th row from the top row. Each number in this row represents a certain combination:

		rows
the 1 represents the empty set	1	0
the 5 represents 5 objects taken 1 at a time	1 1	1
the 10 represents 5 objects taken 2 at a time	1 2 1	2
the 10 represents 5 objects taken 3 at a time	1 3 3 1	3
the 5 represents 5 objects taken 4 at a time	1 4 6 4 1	4
the 1 represents 5 objects taken 5 at a time	1 5 10 10 5 1	5

Thus, the possible combinations of 5 objects taken 3 at a time are 10.

Another example of the use of the Pascal's triangle in **probability** deals with coins. For example, toss 4 coins in the air and list the possible head tail outcomes:

combinations	outcomes
4 heads -HHHH	=1
3 heads & a tail-HHHT,HHTH,HTHH,THHH	=4
2 heads & 2 tails-HHTT,HTHT,HTTH,THTH,TTHH,THHT	=6
1 head & 3 tails-HTTT,THTT,TTHT,TTTH	=4
4 tails-TTTT	=1
total number of outcomes=	16

Note the fourth row of the Pascal triangle gives these outcome numbers. Thus the probability of getting 2 heads when 4 coins are tossed is:

$$\text{probability} = \frac{\text{possible number of outcomes with two heads}}{\text{total number of outcomes}} = \frac{6}{16}$$

Notice that **16** is the sum of the numbers in the 4th row. Also note, the sum of the numbers in any row of the Pascal triangle would be 2^n, where n represents the nth row. Thus the sum of the numbers in the 4th row is $2^4 = 16$ and the sum for the 9th row would be 2^9. This formula can be derived by noting that each row's sum is twice the preceeding row's sum.

```
      1       <---- 2^0=1
   1     1    <---- 2(2^0)=2^1
1     2     1 <---- 2(2^1)=2^2
```

II. Fibonacci sequence & the golden ratio

In the chapter on the golden ratio, it is pointed out that the sequence of consecutive Fibonacci ratios,

1/1, 2/1, 3/2.,5/3, 8/5, . . . ,F_{n+1}/F_n,
1, 2, 1.5, 1.6, 1.625, 1.6153, 1.619, . . .

alternates above and below the value of the golden ratio, ϕ , The limit of this sequence is ϕ. This connection implies that whenever the golden ratio appears the Fibonacci sequence is present and vice versa. (For additional information see the lesson on the golden rectangle.)

III. the Fibonacci sequence & nature

The occurence of the Fibonacci sequence in nature is so frequent that one is convinced it cannot be accidental.

a) Consider the list of the following *flowers with a Fibonacci number of petals* : trillium, wildrose, bloodroot, cosmos, buttercup, columbine, lily blossum, iris.

b) Consider those *flowers with a Fibonacci number of petal-like parts:* aster, cosmos, daisy, gaillardia.

these Fibonacci numbers are frequently associated with the following flowers:

3.................lilies' and irises' petals
5.................columbines', buttercups', and larkspurs' petals
8..................delphiniums' petals
13................corn marigolds' petals
21.................asters' petals
34, 55, 84.....daisies' petals'

BLOODROOT

TRILLIUM

COSMOS

WILD ROSE

c) Fibonacci numbers are also found in the ***arrangement of leaves, twigs, and stems.*** For example, select a leaf on a stem and assign it the number 0, then count the number of leaves (assuming none have been broken-off) until you reach one directly in line with the 0-leaf. The total number of leaves should be a Fibonacci number.

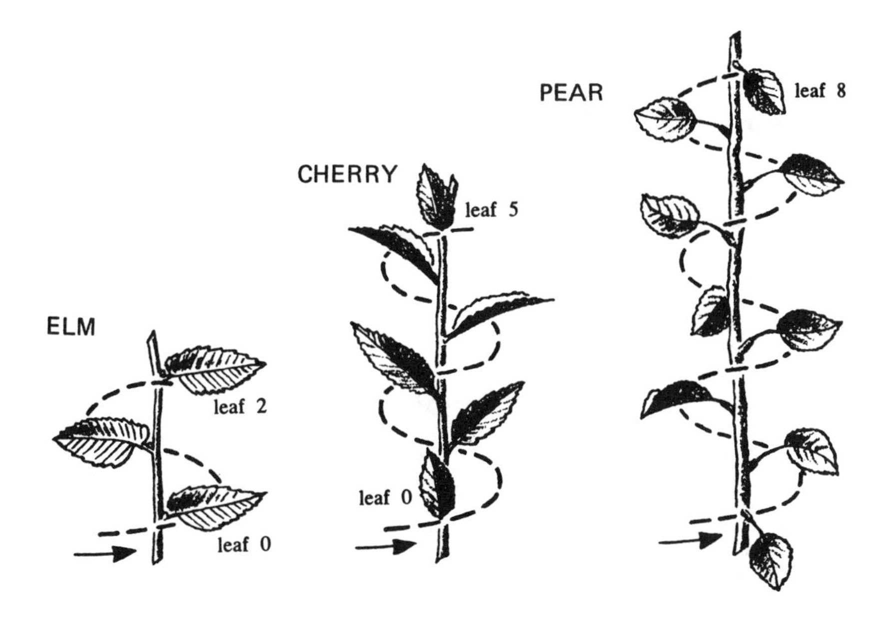

d) Fibonacci numbers have sometimes been called the ***pine cone numbers*** because consecutive Fibonacci numbers have a tendency to appear as left and right sided spirals of a pine cone. This is also true for a sunflower seedhead. In addition, you may find some that are consecutive Lucas numbers.[3]

8 spirals to the right and 13 spirals to the left

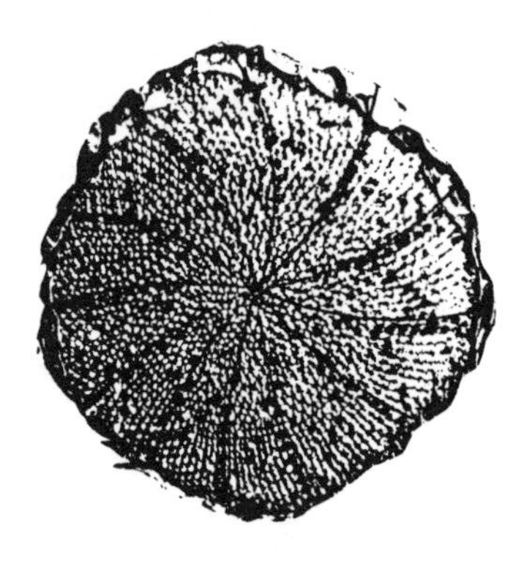

sunflower seedhead

[3] Lucas numbers form a Fibonacci-like sequence which starts with the numbers 1 and 3; then consecutive numbers are obtained by adding the previous two numbers. Thus, the Lucas sequence is 1,3,4,7,11, . . .It is named after Edouard Lucas, the 19th century mathematician who gave the Fibonacci sequence its name and who studied recurrent sequences. Another way in which the Lucas sequence is related to the Fibonacci sequence is the following:

0,1,1,2,3,5,8,13,. . .

1, 3, 4, 7, 11, 18,. . .

e) The ***pineapple*** is another plant to check for Fibonacci numbers. For the pineapple, count the number of spirals composed of hexagonal shaped scales. Look at the base of the pineapple to the identify right and left spirals.

IV. Fibonacci sequence & mathematical tricks

The Fibonacci sequence is linked to some fascinating **"magic"** tricks.

a) Start with any two numbers and begin to generate a Fibonacci-like sequence by adding the preceding two numbers. When you have generated ten numbers in the sequence, stop, and quickly write the sum on a piece of paper and fold it up. You will immediately know the sum, since it is 11 times the 7th term generated.

proof:
1st term=a
2nd term=b
3rd term=a+b
4th term=a+2b
5th term=2a+3b
6th term=3a+5b
7th term=5a+8b
8th term= 8a+13b
9th term=13a+21b
10th term=21a+34b
sum of terms=55a+88b which is 11 times the 7th term:
11(5a+8b)=55a+88b

b) Have someone begin with any two numbers (positive, negative, rational, etc.), and generate as many terms as the person wants, by adding the two preceeding numbers. Then ask someone else to draw a line between any two numbers generated. Now sum the terms above the line in your head by looking at the second number below the line and subtracting from it the 2nd number of the sequence generated. example:

1
5
6
11
__
17 28-5=23 which is the sum of 1+5+6+11
28
45

LESSON

As mentioned in the background material, 1, 1, 2, 3, 5, 8, 13, 21, 34, . . . sequence of numbers had first appeared in print in 1202. But it was not until the 19th century in a mathematical recreation book that the sequence was first called Fibonacci numbers. The sequence appears in nature, art, architecture, science, probability, "magic trick", poetry[4], besides mathematical applications.[5]

Guide the students into discovering how the sequence is generated. Once this is accomplished, it is an opportune time to explain how the sequence originated–see background material. Present the Fibonacci rabbit problem as mentioned in Fibonacci's *Liber Abaci.*

Students are always intrigued to discover how the Fibonacci sequence fits into the realm of nature. One way to introduce this is with a box of pine cones (or having previously asked each student to bring a pine cone to class). If pine cones are not available make a transparency from the included illustrations. Demonstrate how to count left and right swelling spirals. Have them each count and write down the number of left and right spirals their pine cone has. See if they discover that these are consecutive Fibonacci numbers.

Now, go to the illustration of the sunflower seedhead (unless you were actually able to get a seedhead). Demonstrate the same property as the pine cone. Emphasize that they do *not* always form in Fibonacci numbers. (You may want to mention Lucas numbers, from the background material, or leave them out completely.)

With flowers, petals, and stems , continue presenting how and where Fibonacci numbers appear in nature by referring to the background material for details. Bring in any flowers or stems that demonstarte the various occurences of Fibonacci numbers. It always has more impact to have physical examples to share with a class.

At this point it would be fun to do a Fibonacci "magic" trick. Refer to background material for *summing the first ten numbers of a Fibonacci-like sequence.*

[4]George Ekle Duckworth, professor of classics at Princeton University, mentions in his book, *Structural Patterns and Proportions in Virgil's Aeneid,* that the FIbonacci sequence was used by Virgil and other Roman poets.

[5]The use of the Fibonacci sequence is even permeating the computer science field. There are applications of the sequence in sorting data, information retrieval, approximating maxima and minima values of functions whose derivatives are unknown, and generating random numbers.

Depending on the mathematical level of your class, (recall) they can select any two numbers to begin their sequence. If they need practice or review on signed numbers, fractions, or decimals here is an ideal opportunity to introduce them into the lesson.

Now have your class discover how the Fibonacci sequence appears in art, architecture, and other forms of nature. Have them study the ratio of consecutive Fibonacci terms below:

$$1/1,\ 2/1,\ 3/2.,5/3,\ \ 8/5, \ldots, F_{n+1}/F_n,$$
$$1,\ \ 2,\ \ 1.5,\ \ 1.6,\ \ 1.625,\ 1.6153,\ 1.619, \ldots \text{------} \phi$$

Point out how the sequence appears to alternate above and below a particular number. Indicate that this is indeed the case and that the number is the *golden ratio,* also known as the *golden mean.* (For additional background information see the golden rectangle lesson.)

Briefly explain how the golden ratio forms the golden rectangle.

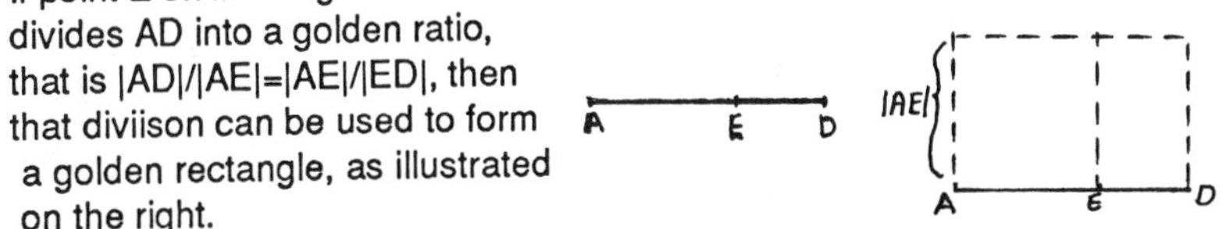

If point E on line segment AD divides AD into a golden ratio, that is |AD|/|AE|=|AE|/|ED|, then that diviison can be used to form a golden rectangle, as illustrated on the right.

Since the limit of the ratio of consecutive Fibonacci numbers, F_{n+1}/F_n, is ϕ, then wherever the golden ratio appears in nature, the Fibonacci sequence is also present. You may want to mention a few objects in art, architecture and nature where the golden ratio appears, such as the Parthenon, the chambered nautilus, some paintings by Leonardo da Vinci, Salvador Dali, George Seurat, Pietter Mondrian. When you do the lesson on the golden rectangle these can be covered in more detail.

You will need to assess your time and your class' level as to whether you will want to present material on the Pascal triangle from the background information.

If you have a beginning math class, you may want to show how the Pascal triangle is formed, give some historical background information, and show where the Fibonacci sequence appears in it. If it is an algebra class, it would be fascinating to show how the Pascal triangle ties into algebra with the binomial expansion–$(a+b)^n$; and link this to probability as mentioned in the background information.

I suggest you end your lesson with a summary and the second Fibonacci "magic " trick as described in the background material.

LESSON'S OBJECTIVES

1) Student will learn the Fibonacci sequence and know how to generate it.

2) Student will be able to generate other Fibonacci-like sequences.

3) Student will learn the historical background of the FIbonacci sequence.

4) Student will discover how the Fibonacci sequence appears in nature.

5) Student will see the link between the Fibonacci sequence and the golden mean.

6) Student will see examples of the Fibonacci sequence in art, architecture, nature through the sequence's link to the golden mean.

7) Student will practice number operations by generating different Fibonacci-like sequences when working "magic" tricks.

8) *Optional*– Student will be introduced to the Pascal triangle, will learn how it is formed, and will learn how it is linked to the binomial expansion and probability.

LESSON'S OUTLINE

I. Introduce the Fibonacci sequence
- A. Discover how it is generated
- B. Historical background
  - 1) Fibonacci's life
  - 2) Fibonacci's book, *Liber Abaci*
  - 3) Edouard Lucas–math recreation book–naming the Fibonacci sequence
    - a) the rabbit problem

II. Fibonacci sequence and nature
- A. Pine cone
  - 1) left and right spirals–consecutive Fibonacci numbers
- B. Sunflower seedhead
- C. Fibonacci numbers in flowers, petals, and stems
- D. Fibonacci "magic" trick
- F. Fibonacci sequence and the golden ratio
  - 1) ratio of consecutive terms, limit and ø

2) Fibonacci sequence and the golden rectangle
   a) golden rectangle in art, architecture, and nature

III. Fibonacci sequence and the Pascal triangle
   A. Formation of the Pascal triangle
   B. Fibonacci sequence in the Pascal triangle
   C. Binomial formula in the Pascal triangle
      1) probability and the Pascal triangle

IV. Summary
   A. review of important points covered
   B. second Fibonacci "magic" trick

V. Pass out assignment.

ANSWERS to the Fibonacci assignment

1)
a) 1,1,2,3,5,8,13,21,34,55,89,144,233,377,610
b) 3,4,7,11,18,29,47,76,123,199,322,521
c) yes (it is the Lucas sequence)

2) number of left spirals=13 number of right spirals=8

3) 3,5,8,8

4) (a)2 (b)4 (c)7 (d) 12
(e) To find the sum the first 92 Fibonacci terms, take the 94th term and subtract 1 from it. To find the sum the first n Fibonacci terms, take $F_{n+2} - 1$.

5) (a) 3 (b) 8 (c) 21 (d) yes yes--------To find the sum of the odd numbered terms to the 19th Fibonacci term, i.e. F_{19}, find the F_{20} and this will be the sum.
In general, $F_1 + F_3 + F_5 + \ldots + F_{2n-1} = F_{2n}$

6) 1,1,2,3,5,8 appear as total of horizontal rows of leaves

ASSIGNMENT

1)

a) List the first 15 Fibonacci numbers______________________

b) Now form a new sequence by doing the following procedure:
add the 1st and the 3rd, the 2nd and the 4th, 3rd and the 5th, 4th and the 6th, 5th and 7th, 6th and 8th, 7th and 9th, 8th and 10th, 9th and 11th, 10th and 12th, 11th and 13th, 12th and 14th, 13th and 15th,. . .

c) Is the sequence formed in part (b) a Fibonacci-like sequence (that is, is each term the sum of the two preceeding terms)?

2) Count the number of left and right spirals in this seedhead.

number of left spirals=_____ number of right spirals=_____

3) Count the number of petals in each flower below, and write your answer under the flower. These particular flowers usually have a Fibonacci number of petals. You may want to find some specimens and check them.

BLOODROOT

WILD ROSE

TRILLIUM

COSMOS

4) Referring to your list in problem 1(a) find the following:
a) sum the first two Fibonacci numbers__________
b) sum the first three Fibonacci numbers__________
c) sum the first four Fibonacci numbers__________
d) sum the first five Fibonacci numbers__________

Do you see these sums appearing as numbers in the Fibonacci sequence?

____

e) Make up a general rule, in your own words, that would describe how to sum the first 92 Fibonacci numbers

____

____

5) Again refer to your list in problem 1(a).
a) sum the 1st and 3rd Fibonacci numbers__________
b) sum the 1st, the 3rd, and the 5th Fibonacci numbers __________
c) sum the 1st, the 3rd, the 5th, and the 7th Fibonacci numbers
d) Do these appear in the Fibonacci sequence?__________

If so, make up some rule that would sum the 1st+3rd+5th+7th+9th+11th+13th+15th+19th Fibonacci numbers.__________

____

e) Find the sum of the first 19 odd Fibonacci terms, that is find 1st +3rd+ 5th+. . .+17th +19th =__________

6) Another way in which Fibonacci numbers often appear in nature is shown in the diagram below:

Total the leaves in the horizontal rows indicated by dotted lines.

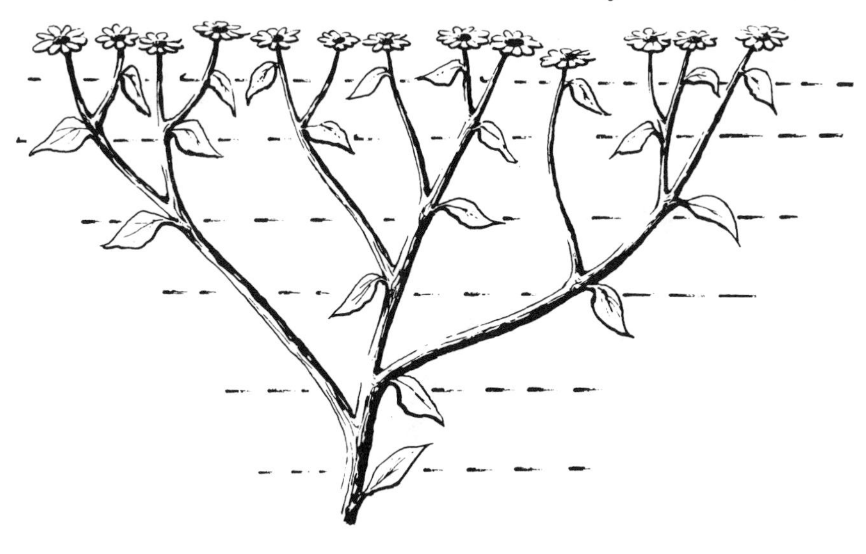

THE GOLDEN RECTANGLE

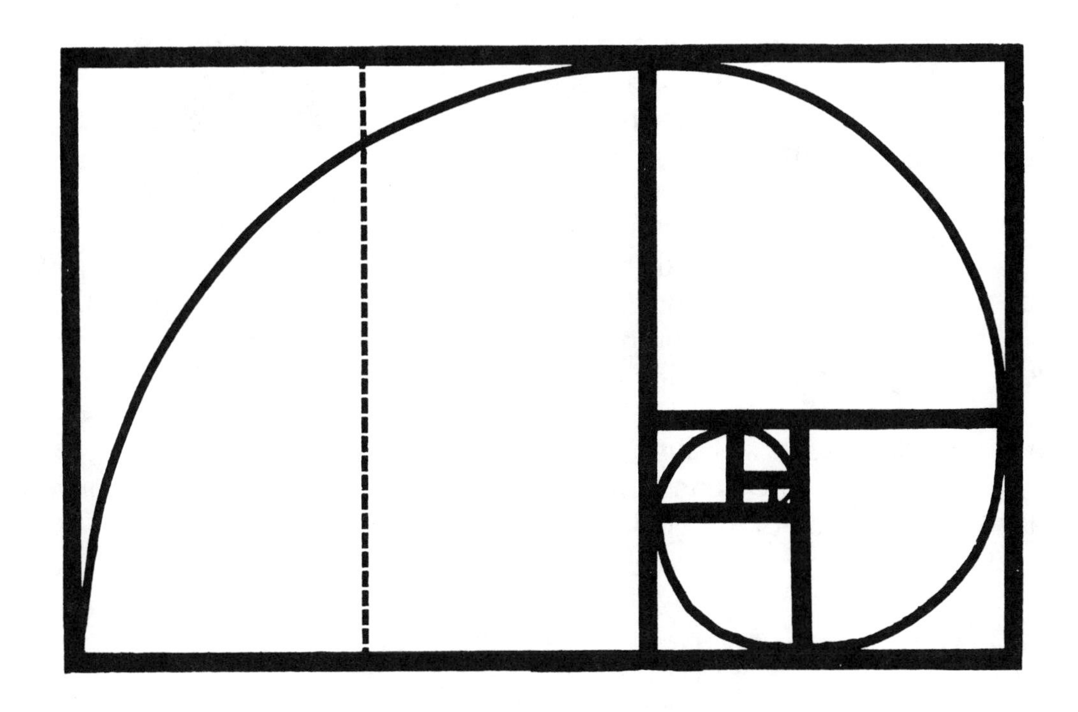

THE GOLDEN RECTANGLE

BACKGROUND MATERIAL

The golden rectangle is a very beautiful and exciting mathematical object, which extends beyond the mathematical realm. Found in art, architecture, nature, and even advertising, its popularity is not an accident. Psychological tests have shown the golden rectangle to be one of the rectangles most pleasing to the human eye.

Ancient Greek architects of the 5th century B.C. were aware of its harmonious influence. The Parthenon is an example of the early architectural use of the golden rectangle. The ancient Greeks had knowledge of the golden mean, how to construct it, how to approximate it[1], and how to use it to construct the golden rectangle. The *golden mean* ,ϕ (phi), was not coincidently the first three letters of *Phidias,* the famous Greek sculptor. Phidias was believed to have used the golden mean and the golden rectangle in his works. The society of Pythagoreans may have chosen the pentagram as a symbol of their order because of its relation to the golden mean.

Besides influencing architecture, the golden rectangle also appears in art. In the 1509 treatise *De Divina Proportione* by Luca Pacioli, Leonardo da Vinci illustrated the golden mean in the make up of the human body. The use of the golden mean in art has come to be labeled as the technique of *dynamic symmetry.* Albrecht Dürer, George Seurat, Pietter Mondrian, Leonardo da Vinci, Salvador Dali, George Bellows all used the golden rectangle in some of their works to create dynamic symmetry.

THE GOLDEN MEAN and THE GOLDEN RECTANGLE

Before discussing the golden mean, we need to define geometric mean or mean proportion.

[1] The Greeks approximated irrational numbers by forming *a ladder of numbers* – for details see pp. 98-99 volume 1 **World of Mathematics** by John R. Newman, Simon and Schuster, New York, 1956.

Mondrian is said to have approached every canvas in terms of the golden rectangle.

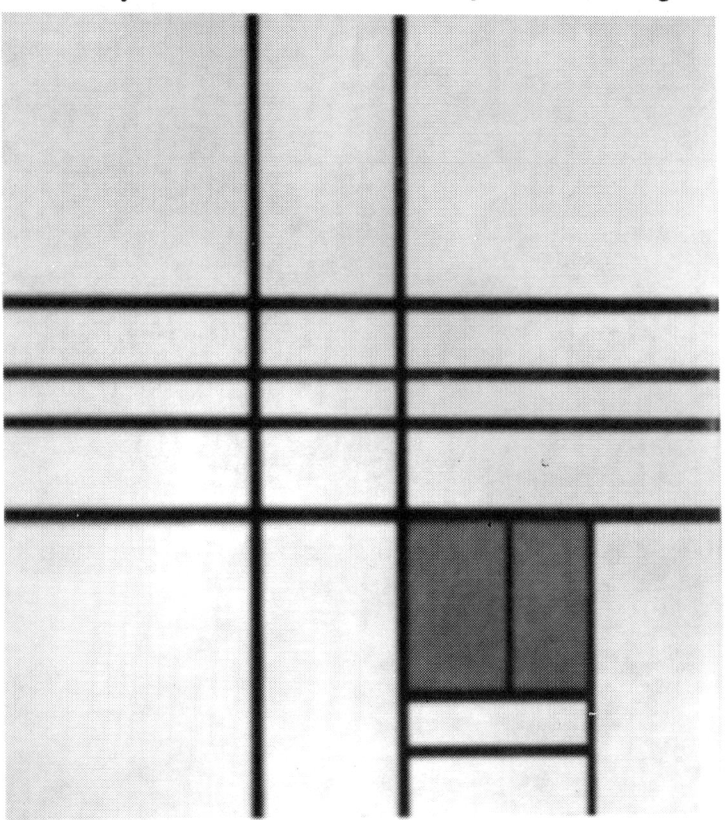

Composition with Yellow, 1936 by Mondrian.

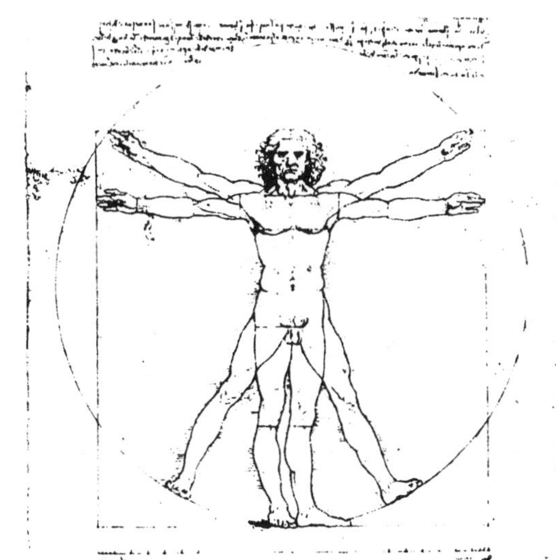

Leonardo da Vinci extensively studied the proportions of the human body. This drawing of his has been studied in detail, and shown to illustrate the use of the golden section. The drawing first appeared in a book he illustrated for mathematician Luca Pacioli called *De Divina Proportione* published in 1509.

The Sacrament of the Last Supper by Salvador Dali (National Gallery of Art, Washington, D.C., the Chester Dale Collection) was printed on a canvas in the shape of a golden rectangle.

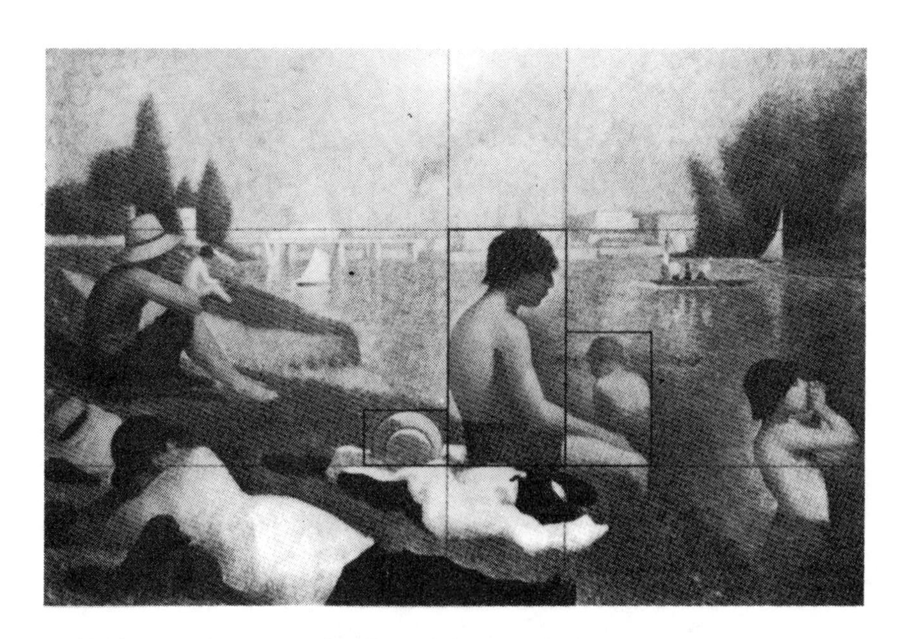

Bathers (1883-1884) by French impressionist George Seurat. There are three golden rectangles shown.

The *mean proportion* is a positive number, x, so that (a/x)=(x/b), and so x is called the *mean proportion* between a and b.

examples of mean propotion in geometry

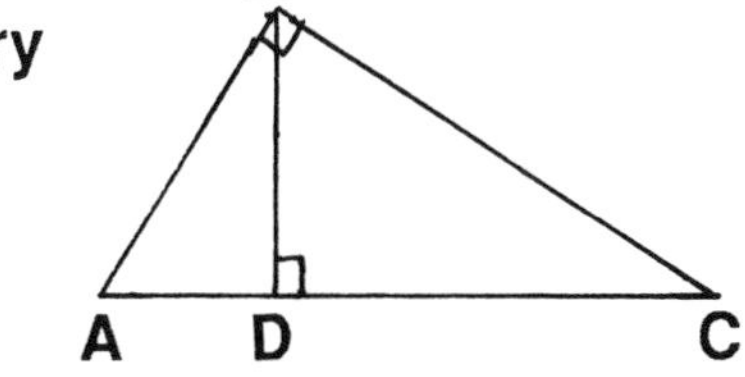

a) altitude to the hypotenuse of a right triangle

since $\Delta ABC \sim \Delta ADB \sim \Delta BDC$
(1) (2) (3)

now using (2) and (3) we get the mean proportion (|BD|/|AD|)=(|DC|/|BD|) so |BD| is the geometric mean;

or using (1) and (2) we get the mean proportion (|AB|/|AC|)=(|AD|/|AB|) so |AB| is the geometric mean here;

or finally using (1) and (3) we get the mean proportion (|BC|/|AC|)=(|DC|/|BC|) so |BC| is the geometric mean.

b) When the geometric mean is located on a given segment, $\overline{AC}$, the golden mean is formed.

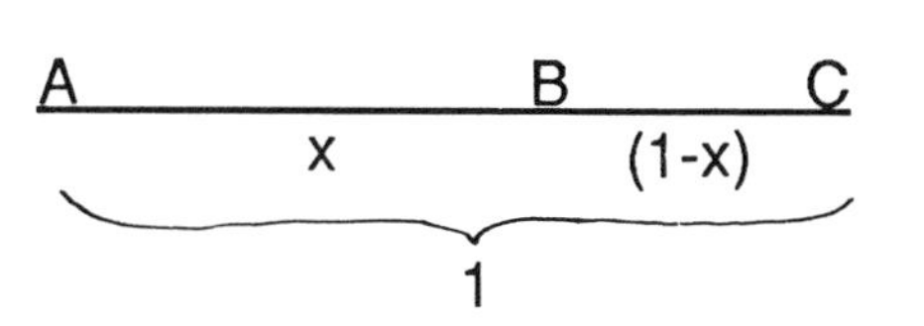

so that (|AC|/|AB|)=(|AB|/|BC|), then |AB| is the *golden mean*, also known as the *golden section*, the *golden ratio*, or the *golden proportion.*

One way to determine the actual value of the golden mean is as follows:

$$1/x = x/(1-x)$$
$$1-x = x^2$$
$$-x^2 - x + 1 = 0$$
$$x = \frac{1+\sqrt{5}}{-2} \text{ or } \frac{1-\sqrt{5}}{-2}$$

Thus the length of $\overline{AB}$ must be represented by the positive value , namely $\frac{1-\sqrt{5}}{-2}$ and the value of the golden mean is its the reciprocal because |AC|/|AB| or |AB|/|BC| represents the golden ratio.

Thus, $|AC|/|AB| = \dfrac{1}{\frac{1-\sqrt{5}}{-2}} \approx \dfrac{1+\sqrt{5}}{2} \approx 1.6$, the golden ratio.

HOW TO DIVIDE A SEGMENT INTO A GOLDEN MEAN

1) Draw segment $\overline{AC}$.
2) Draw $\overline{CD}$ perpendicular to $\overline{AC}$ and |CD|=(1/2) |AC|
3) Draw in AD.

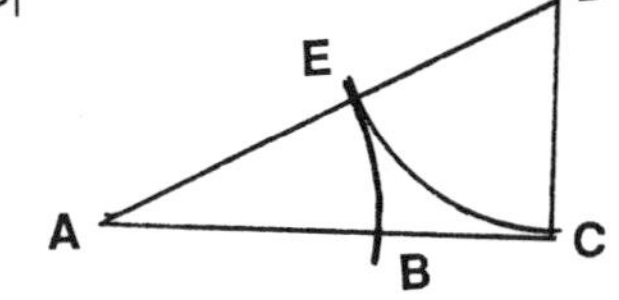

4) With center D and radius |DC|, draw an arc intersecting $\overline{AD}$ at E.
5) With center A and radius |AE|, draw an arc intersecting $\overline{AC}$ at B.
6) B's location divides $\overline{AC}$ into the golden mean.

The proof is similar to the above, and left to the reader.

Once a segment has been divided into a golden mean, the golden rectangle can easily be constructed as follows:

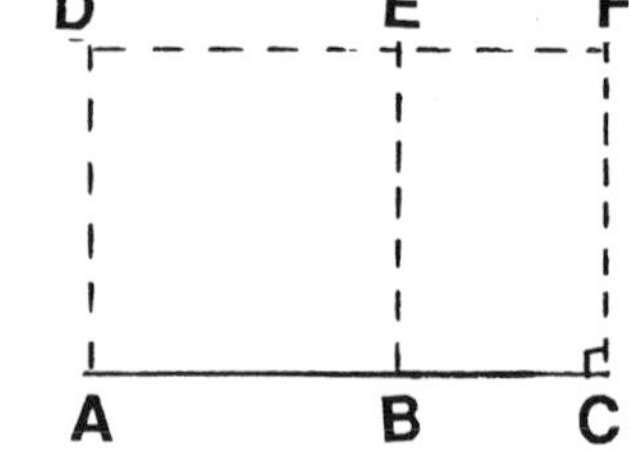

1) Given any segment $\overline{AC}$, with B as its golden mean construct square ABED.
2) Construct $\overline{CF} \perp \overline{AC}$.
3) Extend ray $\overrightarrow{DE}$ so that $\overrightarrow{DE} \cap \overrightarrow{CF} = F$.
Then ADFC is a golden rectangle.

A golden rectangle can also be constructed without already having the golden mean, as follows:

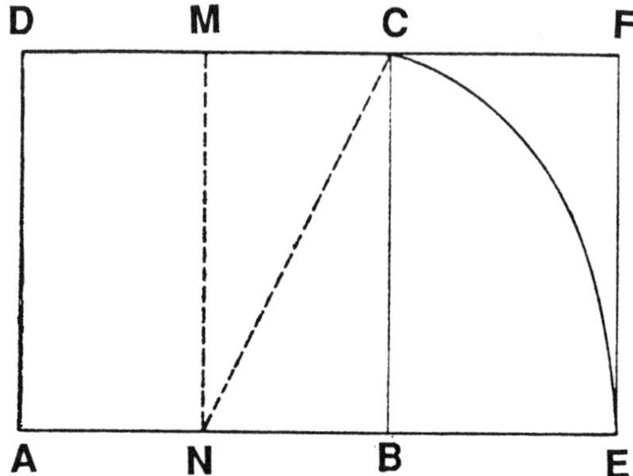

1) Construct any square, ABCD.
2) Bisect the square with $\overline{MN}$.
3) Make arc $\widehat{EC}$ using center N and radius |CN|.
4) Extend ray $\overrightarrow{AB}$ until it intersects the arc at point E.
5) Extend ray $\overrightarrow{DC}$.
6) Construct $\overrightarrow{EF} \perp \overrightarrow{AE}$ and $\overrightarrow{DC} \cap \overrightarrow{EF} = F$.
Then ADFE is a golden rectangle
(See footnote 2 for proof)

The golden rectangle is *self-generating* . Starting with golden rectangle ABCD, golden rectangle ECDF is easily made by drawing square ABEF. Then golden rectangle DGHF is easily formed by drawing square ECGH. This process can be continued indefinitely.

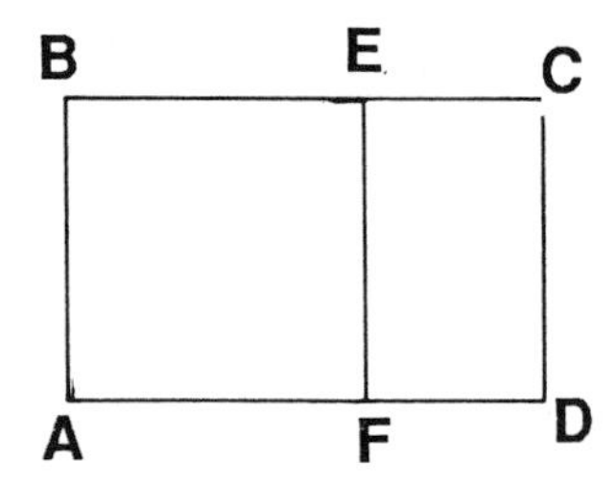

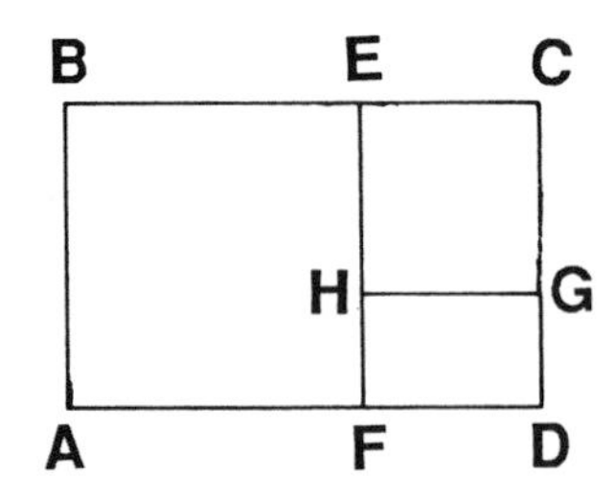

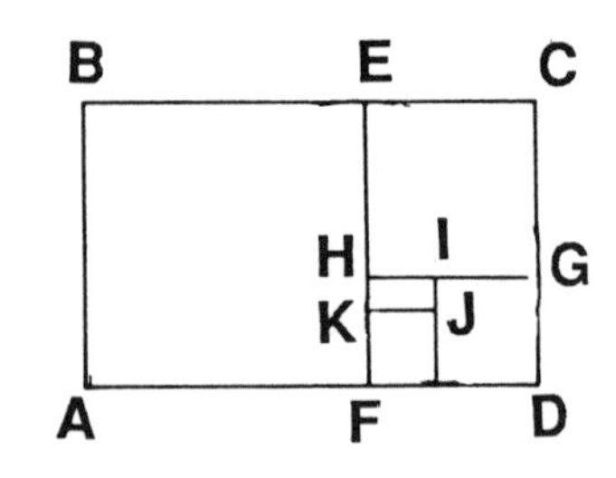

---

2 $y^2 = x^2 + (x/2)^2$

$y^2 = (5x^2)/4$

$y = (x\sqrt{5})/2$

Thus $|BE| = y - (x/2) = [(x\sqrt{5})/2] - (x/2)$

C, y, x, (x/2), A, B, E, x=1

|AE|/|AB| and |AB|/|BE| are equal and both represent φ

$$\varphi = \frac{|AE|}{|AB|} = \frac{x + (x\sqrt{5}/2) - (x/2)}{x} = \frac{x}{(x\sqrt{5})/2 - (x/2)} = 1.618\ldots$$

Using the final product of these infinitely many golden rectangles nestled in one another the *equiangular spiral* (also called the *logarithmic spiral*) can be made. Using a compass and the squares of these golden rectangles, make arcs which are quarter circles of these squares. These arcs form the equiangular spiral.

Note:
The golden rectangle continually generates other golden rectangles and thus outlines the equiangular spiral. The intersection of the diagonals pictured is the ***pole*** *or center of the spiral.*

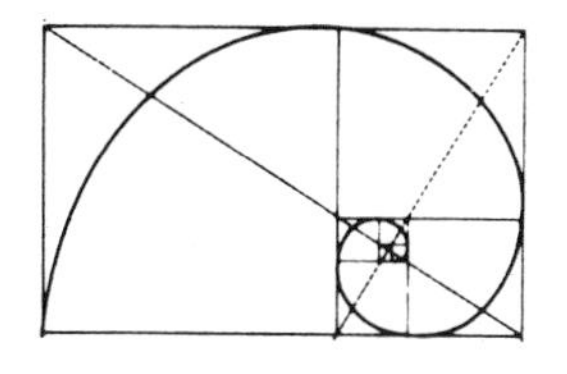

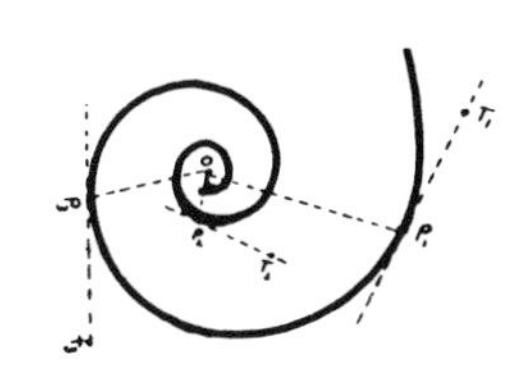

O is the center of the spitral.

A radius of the spiral is a segment with endpoints the center O and any point of the spiral.

Notice that each tangent to the point of the spiral forms an angle with that point's radius, e.g. angle T_1 P_1 O. The spiral is an equiangular spiral if all such angles are congruent.

This is also called a logarithmic spiral because it increases at a geometric rate, .i.e. a power of some number and a power or exponent is another name for logarithm.

The equiangular spiral is the only type of spiral that does not alter its shape as it grows.

Nature has many forms of packaging–squares, hexagons, circles, triangles. The golden rectangle and the equiangular spiral are two of the most aesthically pleasing forms. Evidence of the equiangluar spiral and the golden rectangle are found in starfish, shells, ammonites, the chambered natulilus, seedheed arrangement, pine cones, pineapples, and even the shape of an egg.

Equally exciting is how the golden ratio is linked to the Fibonacci sequence. The limit of the sequence of ratios of consecutive terms of the Fibonacci sequence is the golden mean.

$1/1,\ 2/1,\ 3/2,\ 5/3,\ 8/5,\ 13/8,\ 21/13, \ldots, F_{n+1}/F_n \longrightarrow \phi$

(See the background information on the Fibonacci sequence for additional details.)

The golden rectangle is even used today in advertising and merchandising. Many containers are shaped as golden rectangles to appeal to the public's aestheic point of view.

Yet the golden rectangle interrelates with other mathematical ideas. A few of these are:

1) INFINITE SERIES

The golden mean can be expressed as a sum of infinite series.

$$\phi = 1 + \cfrac{1}{1 + \cfrac{1}{1 + \cfrac{1}{1 + \cdots}}} \qquad \phi = \sqrt{1 + \sqrt{1 + \sqrt{1 + \ldots}}}$$

2) ALGEBRA

The golden mean is linked to the solution of the equation: $x^2 + x - 1 = 0$

3) INSCRIBED REGULAR DECAGON

The golden mean is the ratio of the radius of a circle to the side of an inscribed regular decagon.

4) PLATONIC SOLIDS

The golden rectangle appears in the following Platonic solids:

a) If three golden congruent rectangles intersect each other symmetrically so that each is perpendicular to the other two, then their twelve vertices are the vertices of a regular icosahedron.

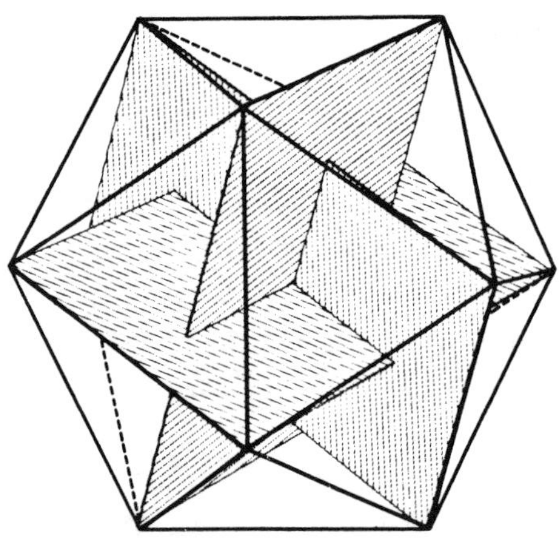

b) The twelve vertices are the centers of the faces of a regular dodecahedron.

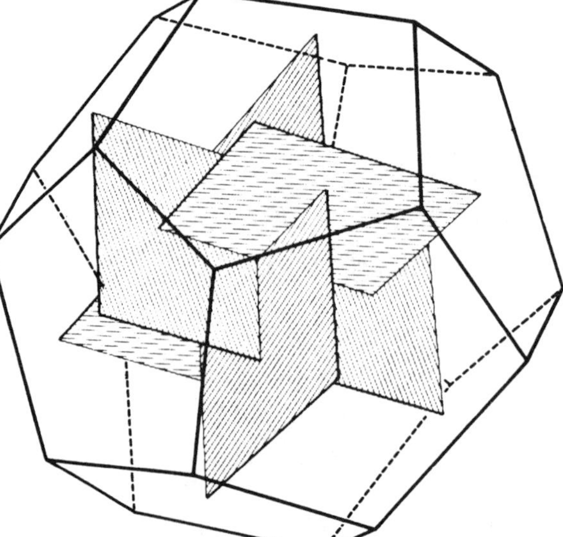

5) EQUIANGULAR SPIRAL & LOGARITHMS SPIRAL

The golden rectangle generates the equiangular spiral, and the diagonals pictured interesect at the pole of the spiral.

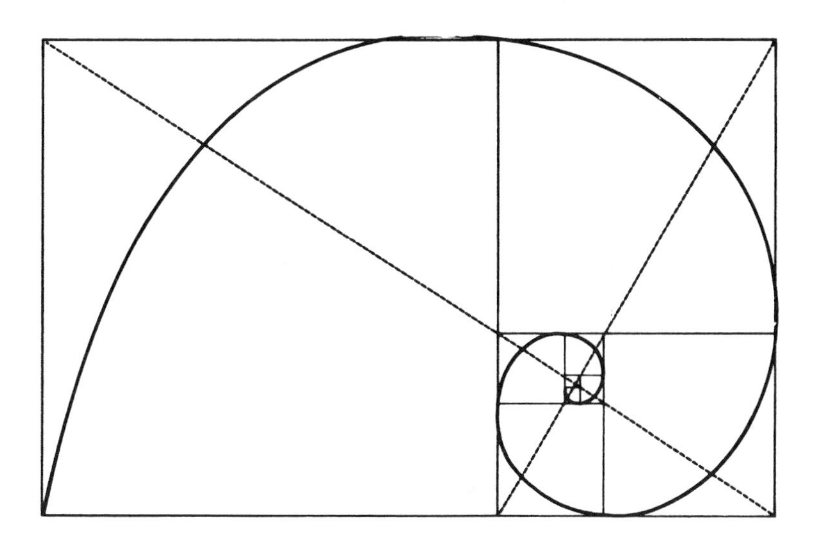

6) LIMITS

The golden mean is the limit of the sequence formed from ratios of consecutive Fibonacci numbers, 1, 1, 2, 3, 5, 8, 13, 21, 34, . . .

1/1, 2/1, 3/2, 5/3, 8/5, 13/8, 21/13, 34/21, . . .

1, 2, 1.5, 1.6, 1.6, 1.625, 1.6153846, 1.6190476, . . . ----------→ ϕ

these terms alternate above and below the value of the golden mean.

7) GOLDEN TRIANGLE

The golden triangle is an isoscles triangle with base angles 72º and vertex angle 36º. As noted from the diagram it is self-generating like the golden rectangle. |AB|:|BC|= ϕ:1. The intersection of the dotted segments is the location of its pole.

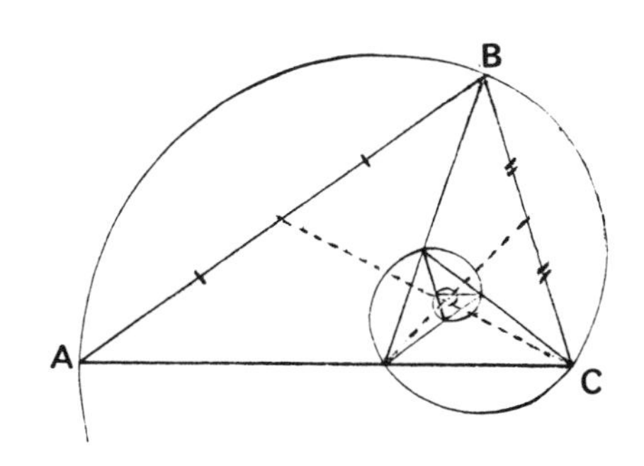

8)PENTAGRAM

The golden mean is related to the regular inscribed pentagon.

Triangles ΔFGE and ΔEBC are golden triangles.

B divides segment AC into the golden mean.
C divides segment BD into the golden mean.

Thus, if |AB|=1, then |BC|=.618. . .

The same is true with other sides of the pentagram, so long as the pentagram is formed from a regular pentagon.

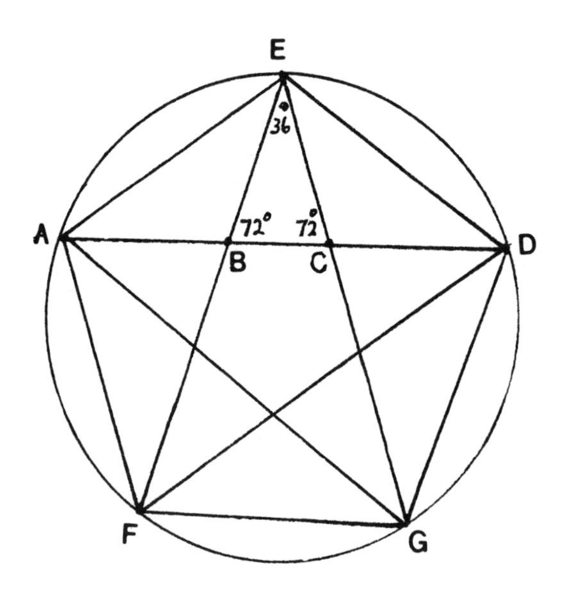

LESSON

Cut out the following shaped rectangles ahead of time or draw them on the board in order to start the lesson with a class survey. Ask your class which is most pleasing aesthically. See if the majority prefer the golden rectangle.

After you have taken your survey, mention that psychological tests have shown that more people prefer the golden rectangle. (The first example is the goldn rectangle.)

With this introduction go into a brief explanation of the golden mean and the golden rectangle, and how they are constructed. It is probably a good idea to mention all the names for the *golden mean – golden ratio, golden proportion,* and *golden section.* See background material information for details and constructions.

The detail of your discussion of the geometric mean and the golden mean will depend on the mathematical level of your class. If it is not an advanced class of at least geometric background you may just want to explain that the golden mean is a special ratio (ratio–a comparison of two quantities, when two ratios are equal it is called a proportion) in which a segment is divided in the following manner:

where B is located on AC so that $(|AB|/|AC|)=(|BC|/|AB|)$ A B C

If your class has covered quadratic equations and the quadratic formula, then it would be exciting to discover with them the value of the golden mean, ϕ,–see background material for details. At this point it might be interesting to mention the ϕ is not as well known as π, but it has a way of popping up in all sorts of places when least expected (one must be familiar with ϕ and its value in order to recognize it). ϕ is one of the values that nature used in the formation of many of its objects, as brought out in the background material.

Referring to background information, construct the golden rectangle with your class. The easiest method to use is the one that starts with a square. Use your judgement as to whether your class' background would lend itself to the proof of why this construction gives the golden mean.

Again referring to the background material and the diagrams provided, illustrate the golden rectangle's presence in:

1) art
2) architecture
3) nature
4) merchandising
5) golden mean's relation to other mathematical concepts

The fact that the golden rectangle is able to generate infinitely many golden rectangles within itself is always fascinating to classes. Two ways to illustrate this are:

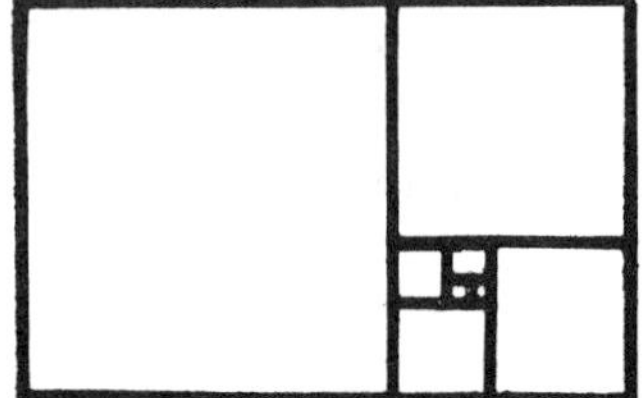

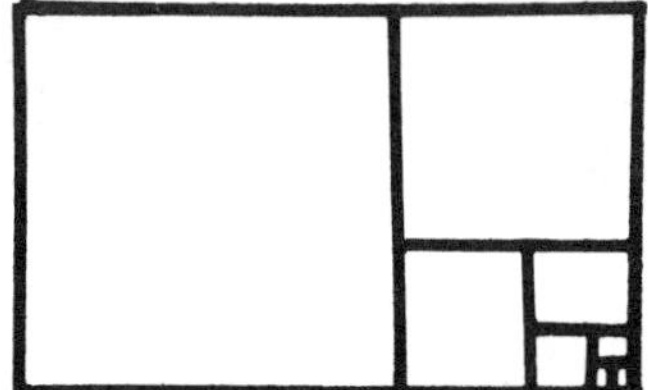

Now using the first construction above of the self-generating golden rectangles, show your class how the equiangular spiral is formed by constructing with your compass quarter circles in the squares. You may want to explain why this is an equiangular (or logarithmic) spiral and that the pole of the spiral is the interesection point of the two diagonals. Details of this information are presented in the background material.

Now point out or present actual objects in nature that are in the form of an equiangular spiral and demonstrate how these are packaged by the golden rectangle – see diagram below.

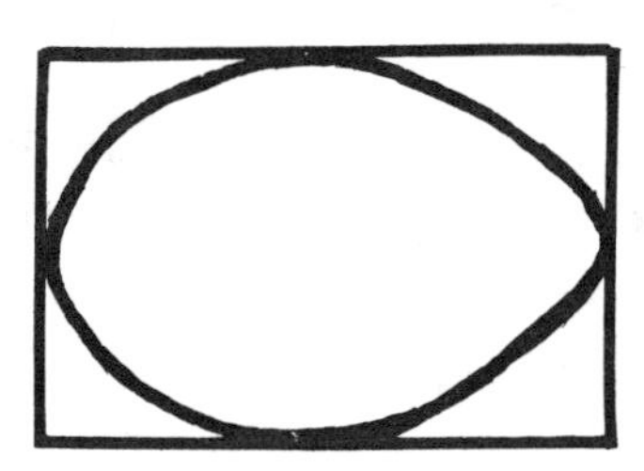

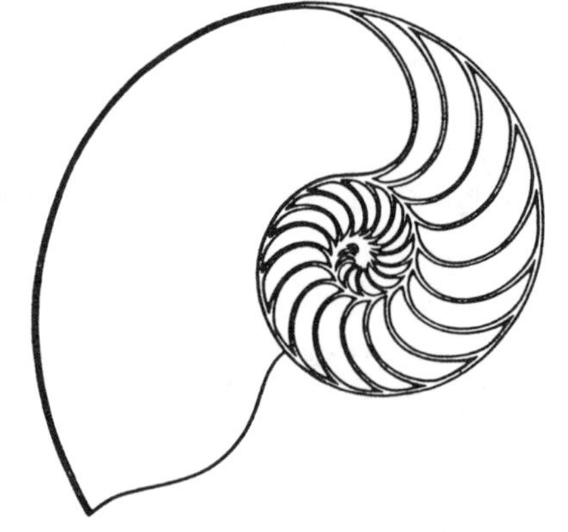

As mentioned in the background material, there are a variety of mathematical concepts that relate to the golden mean and the golden rectangle. You will need to decide which of these are appropriate to discuss with your particular class.

Lastly, if you have done the lesson on the Fibonacci sequence, you can now relate the golden mean or reintroduce it and show its relation to the Fibonacci sequence. As detailed in the background information, the limit of the ratios of consecutive terms of the Fibonacci sequence is the golden mean, that is
1/1, 2/1, 3/2, 5/3, 8/5, 13/8, . . . , F_{n+1}/F_n has limit ϕ =1.618. . .

The above fractions when converted to decimals alternately have values above and below ϕ , and continually come closer to ϕ.

Summarize the lesson and emphasize how amazing it is to see the golden mean and the golden rectangle intermingled in so many aspects of our world.

LESSON'S OUTLINE

I. Introduce the golden rectangle by survey of class
  A. Mention psychological tests

II. Discuss the construction of the golden rectangle and the golden mean.
  A. Mention other names for the golden mean–golden section, golden proportion, golden ratio.
  B. Derive the value of the golden mean.–*optional*
  C. Construct the golden rectangle in class.
  D. Discuss the golden rectangle's presence in other areas.
    1) art
    2) architecture
    3) nature
    4) merchandising
    5) mathematical concepts

III. Demonstrate how the golden rectangle is self-generating.
  A. Equiangular spiral
    1) definition
    2) Using the self-generating golden rectangle, construct the equiangular spiral.
    3) Discuss objects in nature that contain the equiangular spiral and the golden rectangle.

IV. Fibonacci sequence and the golden mean
  A. Demonstrate the sequence for which ø is the limit.

V. Summarize lesson.

VI. Pass out assignment.

LESSON'S OBJECTIVES

1) Student will learn what the golden rectangle and the golden mean are.

2) Student will learn how to construct a golden rectangle.

3) Student will learn how the golden rectangle is self-generating.

4) Student will be introduced to the equiangular spiral, and learn how to form it using a golden rectangle.

5) Student will see the presence of the golden rectangle in art, architecture, nature and merchandising.

optional objectives

6) Student will see how the value of the golden mean is derived.

7) Student will see the proof of the construction of the golden rectangle, and how the value of the golden mean is derived.

8) Student will be introduced to the golden mean's relationship to other mathematical concepts.

ANSWERS to assignment:

1) golden rectangle should be visible
2) See background information
3) See background information
4) Procedure is outlined in detail for students to follow.
5) a) pentagon
   b) pentagram
   c) indefinitely
6) See background material.
7) a) 1, 1.5, 1.666..., 1.6, 1.625, 1.615..., 1.619..., 1.617..., 1.61818...,
   b) 1, 1.414..., 1.553..., 1.598..., 1.611..., 1.616..., 1.617..., . . .

ASSIGNMENT

1) Each object below has a golden rectangle concealed in some way in its shape. Locate the golden rectangle in each and sketch it.

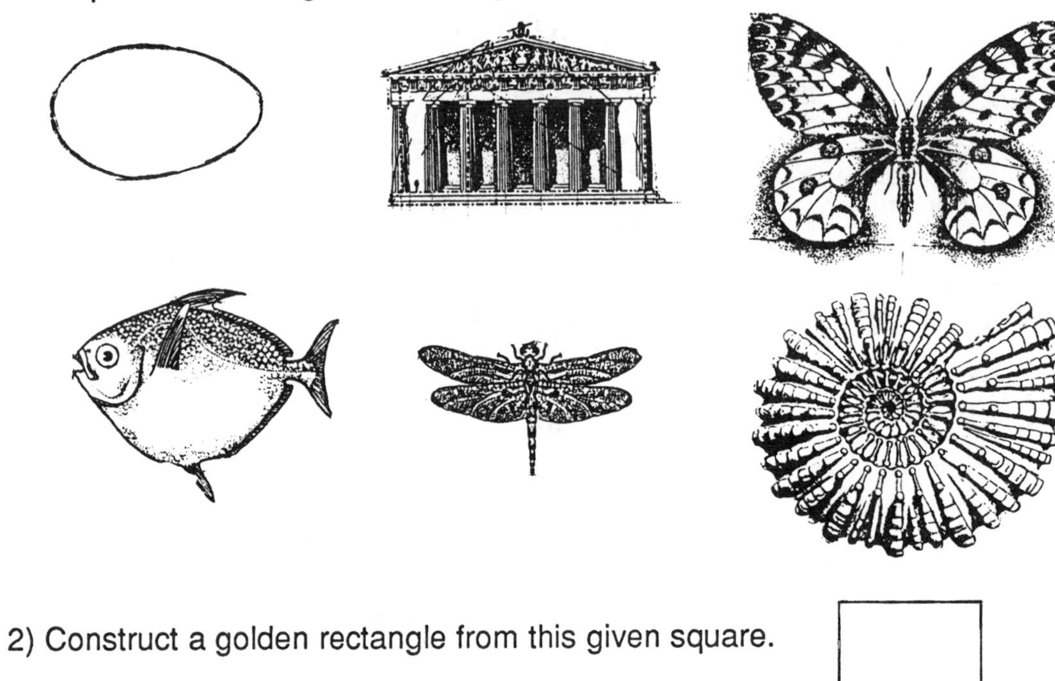

2) Construct a golden rectangle from this given square.

3) Again, construct another golden rectangle from the square below.

   a) Use this golden rectangle to generate 6 more golden rectangles.

   b) Use your work from (a) to from an equiangular spiral.

   c) Draw two diagonals on your equiangular spiral to find its pole– the center of the spiral.

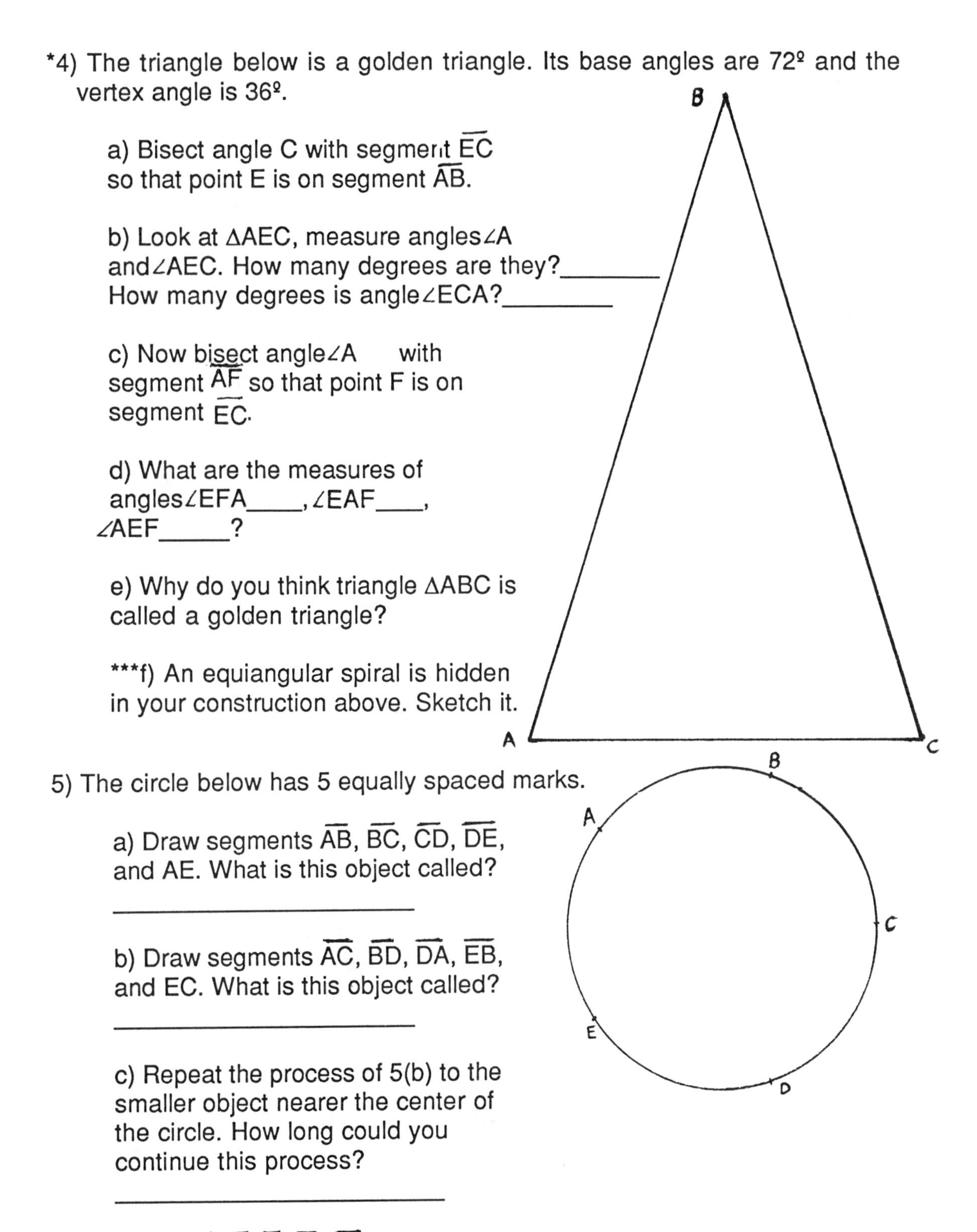

*4) The triangle below is a golden triangle. Its base angles are 72º and the vertex angle is 36º.

a) Bisect angle C with segment $\overline{EC}$ so that point E is on segment $\overline{AB}$.

b) Look at ΔAEC, measure angles $\angle A$ and $\angle AEC$. How many degrees are they?________ How many degrees is angle $\angle ECA$?________

c) Now bisect angle $\angle A$ with segment $\overline{AF}$ so that point F is on segment $\overline{EC}$.

d) What are the measures of angles $\angle EFA$____, $\angle EAF$____, $\angle AEF$_____?

e) Why do you think triangle ΔABC is called a golden triangle?

***f) An equiangular spiral is hidden in your construction above. Sketch it.

5) The circle below has 5 equally spaced marks.

a) Draw segments $\overline{AB}$, $\overline{BC}$, $\overline{CD}$, $\overline{DE}$, and AE. What is this object called?

____

b) Draw segments $\overline{AC}$, $\overline{BD}$, $\overline{DA}$, $\overline{EB}$, and EC. What is this object called?

____

c) Repeat the process of 5(b) to the smaller object nearer the center of the circle. How long could you continue this process?

____

note: segments $\overline{AC}$, $\overline{AD}$, $\overline{BE}$, $\overline{BD}$, $\overline{CE}$ have all been divided into golden means by one of their intersecting points.

6) Assume the rectangle below is a golden rectangle, locate point C on AB so that C divides AB into a golden mean.

***7) 1.618 is a decimal approximation for ϕ . ϕ is an irrational number, so it does not have an exact decimal or fraction value.

a) A method to derive decimal approximations for ø uses infinite fractions.

$$\phi = 1 + \cfrac{1}{1 + \cfrac{1}{1 + \cfrac{1}{1 + \cfrac{1}{1 + \cdots}}}}$$

Consider the approximate values of ϕ , the sequence of numbers derived from the above continued fractions. Calculate the first five values: Give their decimal values.

b) ϕ can also be estimated by the following expression:

Using a hand calculator, find the values of the first five approximations.

$$\phi = \sqrt{1 + \sqrt{1 + \sqrt{1 + \sqrt{1 + \cdots}}}}$$

THE MATHEMATICS OF PAPER FOLDING

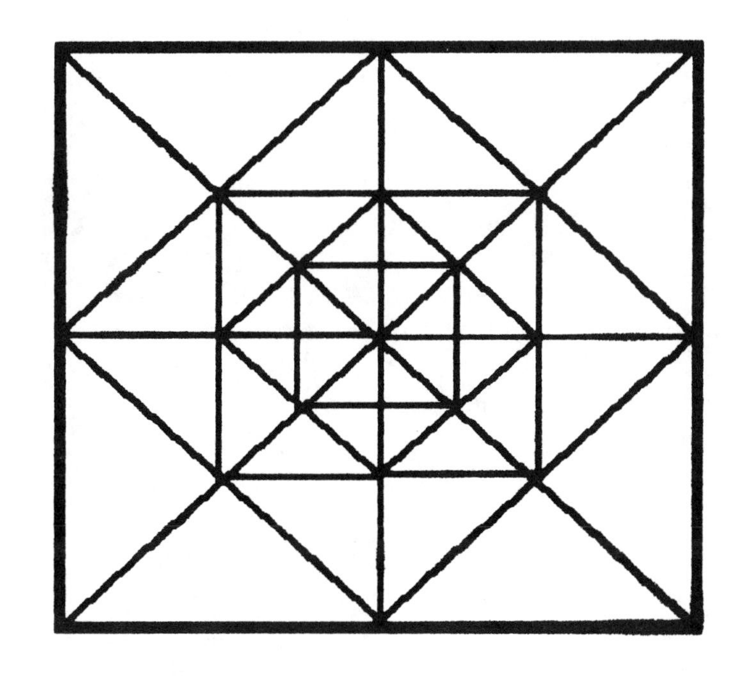

THE MATHEMATICS OF PAPER FOLDING and ORIGAMI

BACKGOUND MATERIAL

Most of us at one time or another have folded a piece of paper and tucked it away, but fewer of us have intentionally folded paper in order to study the mathematical ideas it revealed. Paper folding can be both educational and recreational. Even Lewis Carroll was a paper folding enthusiast. Although paper folding transcends many cultures, it is the Japanese who are associated with developing and popularizing it into the art form called origami.

Buddhist priests brought paper folding methods into Japan from China through Korea in 538 A.D. Since the manufacturing of paper at this time was costly, people used it with care, and origami became part of certain ceremonies. The art of origami has been shared and passed on from generation to generation. Traditional origami, which is the origami used in ceremonies, does not allow the use of scissors, glue, or decorating aids, while modern origami does not have these restrictions. Origami is a balance between the simple and the complex. Beginning with only a square piece of paper, realistic and elaborate animals, flowers, boats, people, or objects can be formed.

MATHEMATICAL ASPECTS OF PAPER FOLDING

1) exponential growth

example: $(1/2)^n$

Take a sheet of newspaper. Fold it in half. Continue the process and note how quickly its size diminishes. Were eight folds even possible?

$(1/2)^8=1/256$ so the resulting rectangle would have to be 1/256 the size of the original piece of paper.

2) geometry

a) Geometric concepts naturally appear when paper folding, giving rise to the following terms.

square, rectangle, right triangle, congruent, diagonal, midpoint, inscribe, area, trapezoid, perpendicular bisector

Here are some examples of the use of these terms:

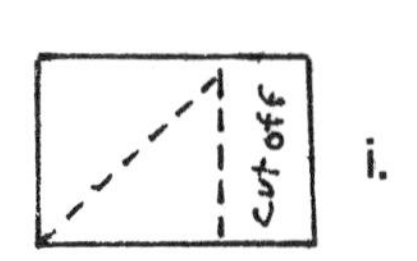

i) from a **rectangular** shaped paper form a **square**

ii) with the square paper form 4 **congruent right triangles**

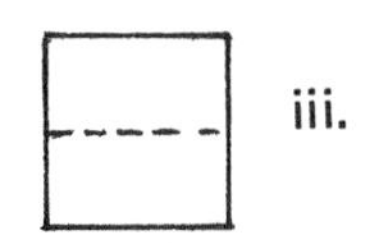

iii) find the **midpoint** of a side of a square

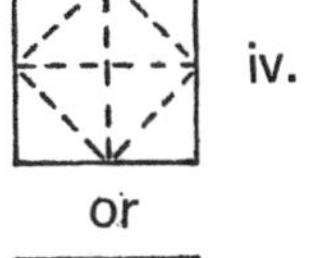

or

iv) **inscribe** a square in the paper square

v) studying the paper's creases, notice that the inscribed square is 1/2 the **area** of the large square

vi) form two congruent **trapezoids** by taking a square sheet of paper and folding it along any edge so the crease passes through the center

vii) make the **perpendicular bisector** of a segment by folding the square in half–the crease will be the perpendicular bisector to the side of the square

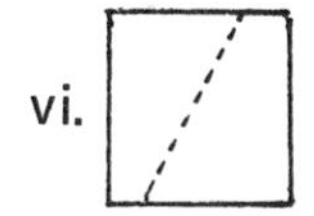

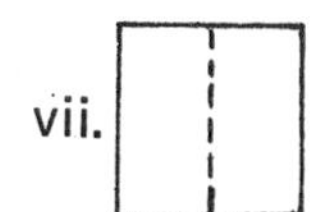

b) Demonstrate the **Pythagorean Theorem**
Fold the square paper as shown in the diagram.

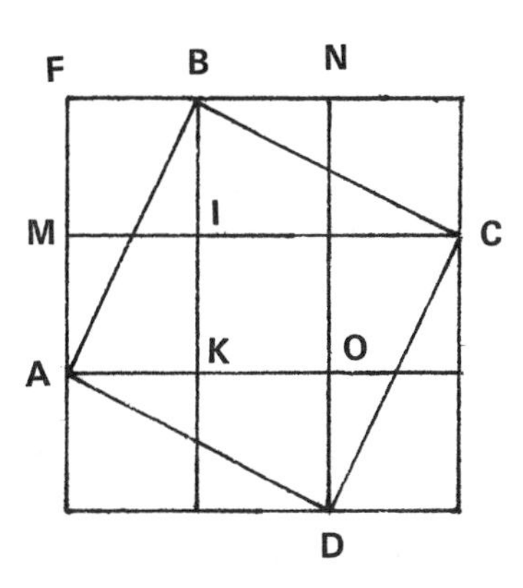

c^2 = area of square ABCD
a^2 = area of square FBIM
b^2= area of square AFNO

By matching up congruent shapes,
the area of square FBIM = area of Δ ABK
and
the area of AFNO = the area of BCDAK
(the remaining area of square ABCD).

Thus, $a^2 + b^2 = c^2$

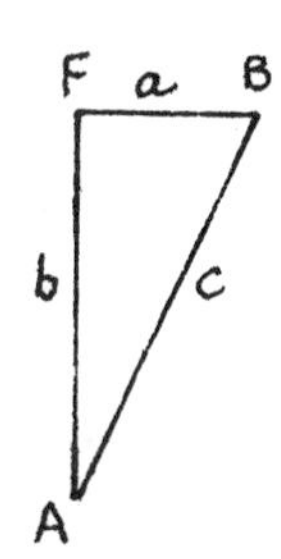

c) Demonstrate the **theorem that the measure of the angles of a triangle total 180 degrees,** by taking any shaped triangle and folding it as illustrated.

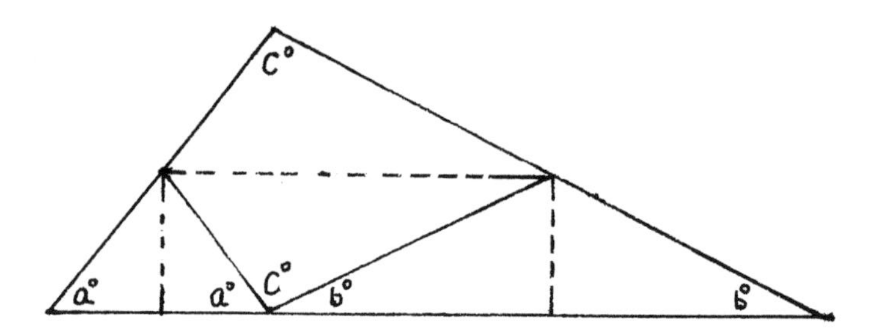

d) Demonstrate **similarity and midpoints** by folding the squares as illustrated.

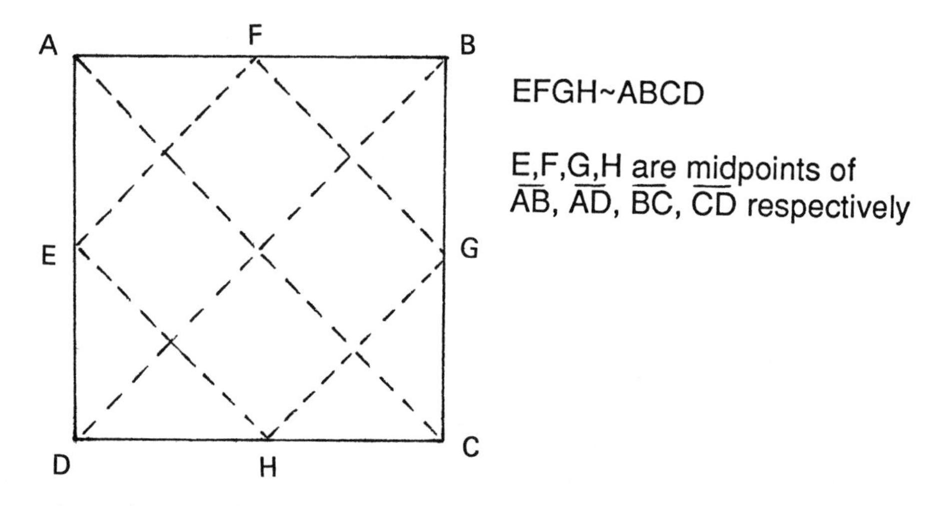

e) Demonstrate that the **angle bisectors of a triangle are concurrent** (intersect in one point) by dividing each angle in half and noting that the three angles bisectors intersect in one point.

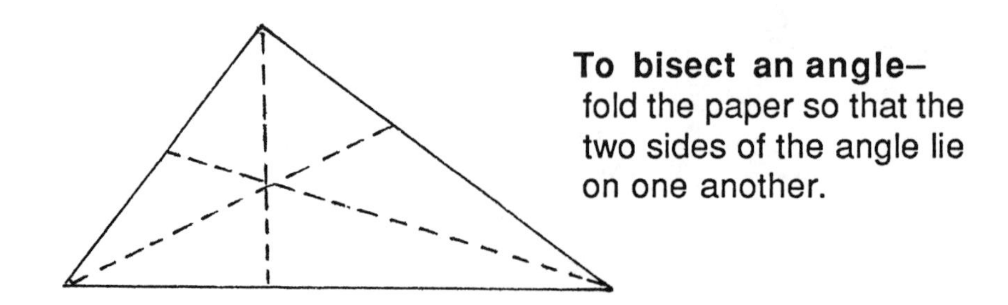

f) Demonstrate that the **medians of a triangle are concurrent** by locating the midpoints of each side, and then make a crease between the midpoint and the vertex of the opposite angle. Do this for each vertex, and note that these medians (creases) all intersect in one point.

g) Construct a **parabola** by folding **tangent lines.**

procedure: Locate the focus point of the parabola a few inches from the side of the sheet of paper. Crease the paper between 20 and 30 times as illustrated in the diagram. These creases are the tangents of the parabola and outline the curve.

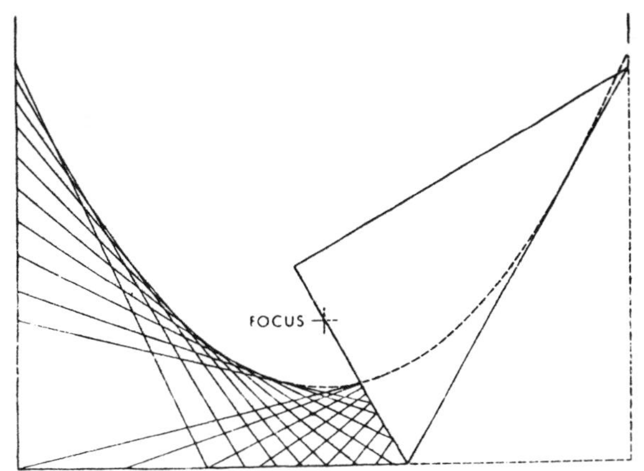

h) Construct a **regular octagon.**

procedure: Inscribe a square as in 2(iv). Bisect angles 1, 2, 3, 4, 5, 6, 7, 8 as illustrated. This process produces the eight congruent sides of the octagon.

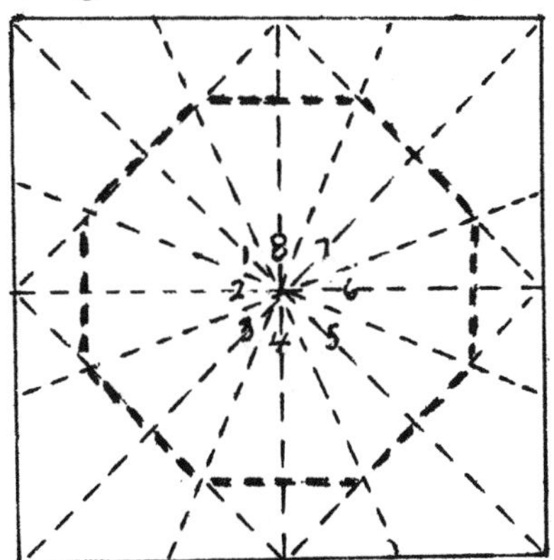

These examples illustrate only some of the mathematical terms and concepts that can be demonstrated using paper folding. In addition, the physical process helps students retain these concepts.

LESSON

Materials needed: (1) one sheet of newspaper (2) a large origami flying bird (see instructions for constructing)

In presenting the lesson, it is important to illustrate from the beginning both aspects of paper folding – the mathematical and the recreational.

An enjoyable way to introduce this lesson is with a **challenge**:

> Take a sheet of newspaper and challenge the class to fold it in half 8 times. Then mention the mathematical principle of exponential growth as presented in the background material.

Now take the large flying bird (see instructions under flying bird) that *you made ahead of time,* and present it to the class. Illustrate how it flaps its wings, and mention it will be one of the origami objects constructed that day.

Give a brief historical background of origami as mentioned in the background material. Emphasize that the origami objects and the mathematical objects which will be constructed will be made by working with a square sheet of paper.

For the following questions encourage class involvement by having students respond to them as you propose them.

procedure:

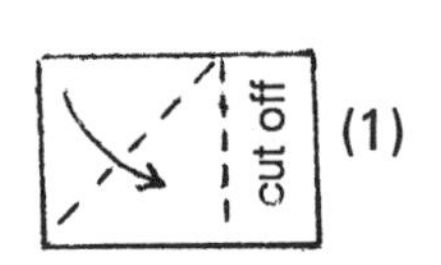

1) Pass out a rectangular sheet of paper to each student, Now ask them to make a square out of it.

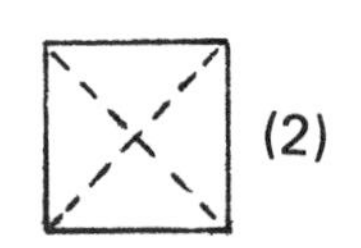

2) Using this square ask them to fold it in order to form 4 congruent right triangle. (Depending on the level of your class you may need to define terms they have not learned, e.g. conguent, right triangle)

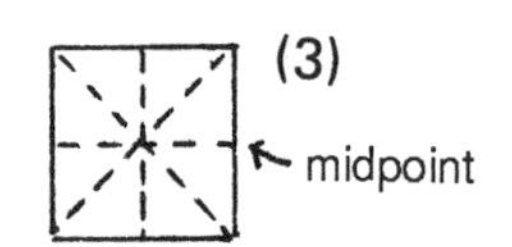

3) With the same square, ask them to figure out a way to fold it in order to find:

   a) the midpoint of the side of a square
   b) a way to divide the square into 8 congruent right triangles

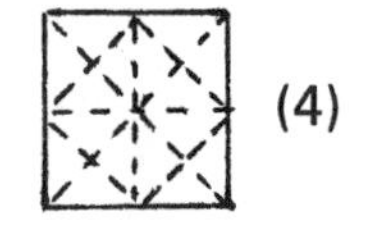

4) Now with the square divided into 8 congruent right triangles, make the appropriate folds so that a square appears inscribed in the original large square.

question for students:

a) How do the areas of the inscribed square and the large square compare?

b) Can this process be continued so another square is formed inscribed in the inscribed square?

c) How many inscribed squares can be formed by this process?

At this point you will have to decide which of the paper folding done in the background material will be best suited for presentation to your particular class needs.

These include: ***Pythagorean Theorem, sum of the measures of the three angles of a triangle, simlarity and midpoint, angle bisectors of a triangle are concurrent, medians of a triangle are concurrent.***

Now, using new square sheets of paper, construct the following origami figures with your class. I recommend the diagrams of the procedures for these be passed out to each student or shown on an overhead projector for them to follow as you do it with them.

1) the box

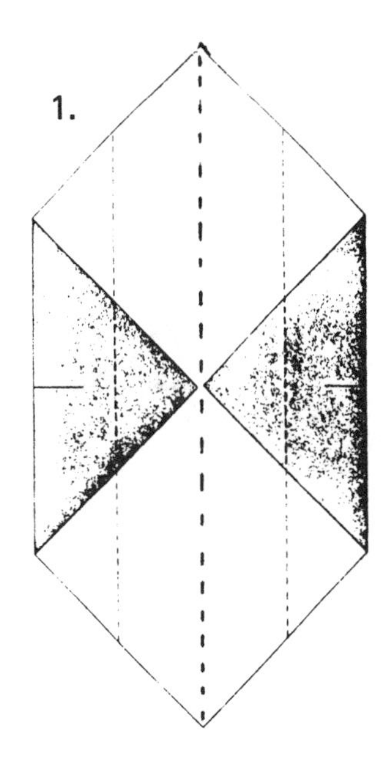

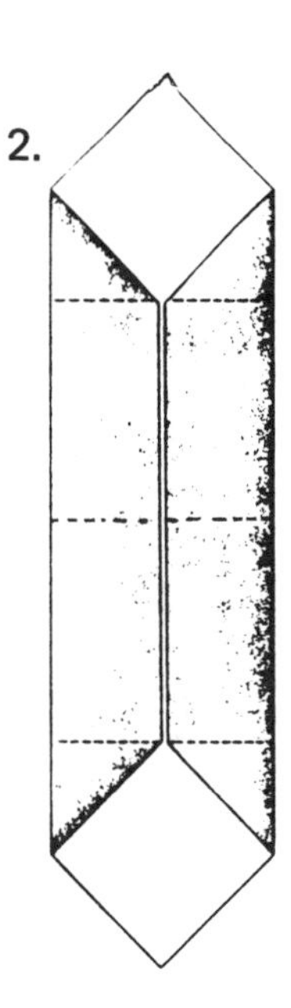

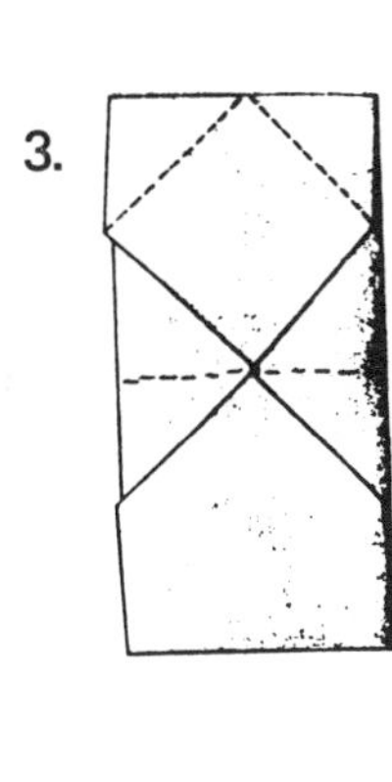

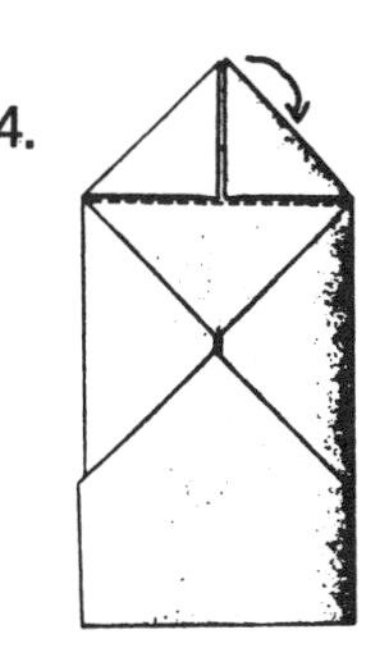

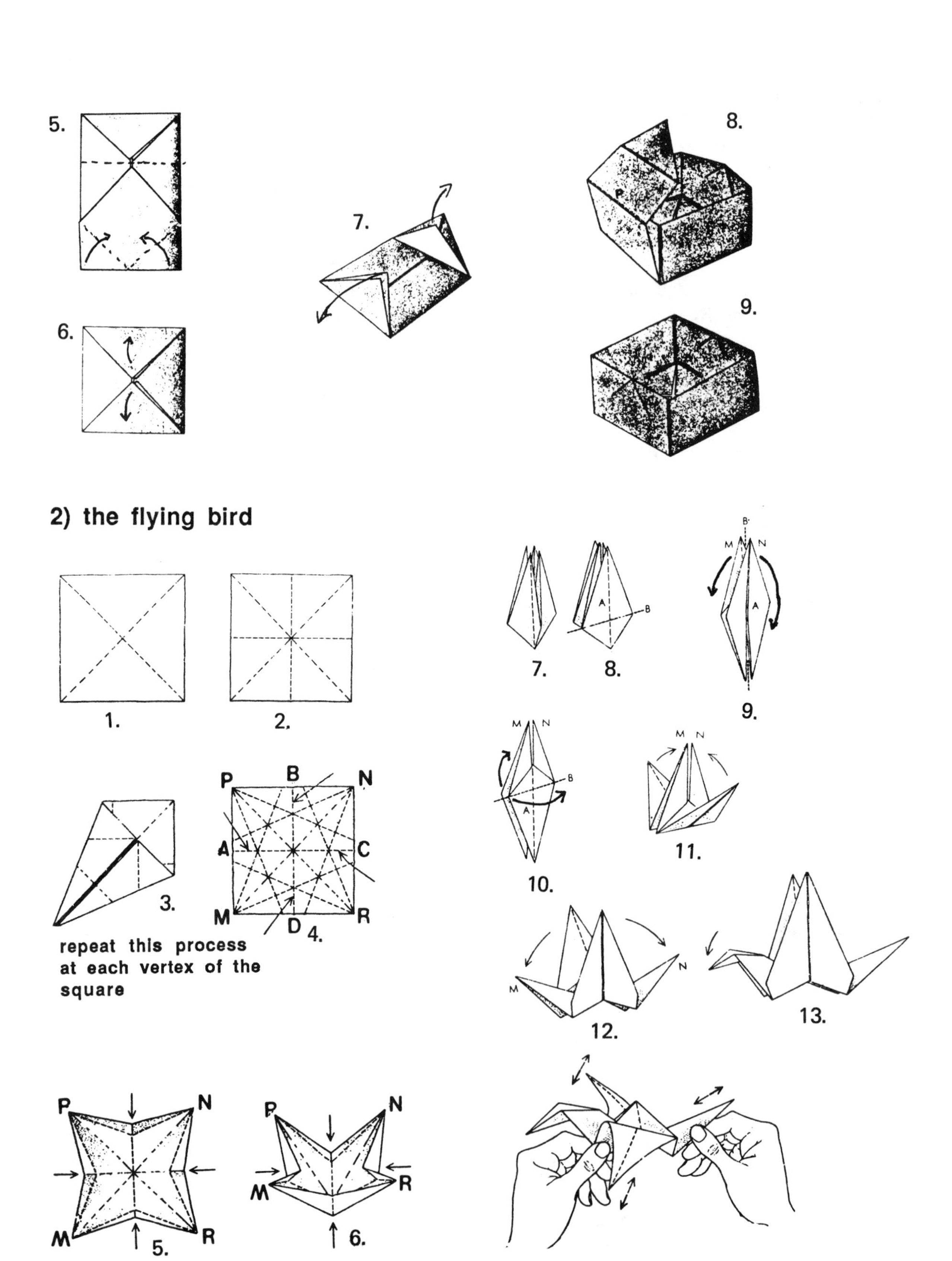
5.
6.
7.
8.
9.
2) the flying bird
1.
2.
3.
P B N
A C
M D R
4.
repeat this process
at each vertex of the
square
P N
M R
5.
P N
M R
6.
A B
7.
8.
B M N A
9.
M N B A
10.
M N
11.
M N
12.
13.

LESSON'S OBJECTIVES

1) The student will learn mathematical terms and concepts, which are described in the background material, through paper folding.

2) The student will be introduced to the art of origami, and will fold various origami figures.

LESSON'S OUTLINE

I. Demonstrate the mathematical and recreational aspects of paper folding.
   A. Challenge problem
      1) an example of exponential growth
   B. Present and demonstrate the origami flying bird

II. Give historical background of origami
   A. Introduction of paper making in Japan
      1) route and manner of introduction
      2) date of introduction
   B. Difference between traditional and modern origami
      1) use of origami in ceremonies
      2) use of glue, scissors and decorating aids

III. Present mathematical aspects of paper folding with students' involvement in the process.
   A. Select those concepts and terms from the background material that are most appropriate to your class' level.

IV. Have students make two origami figures as you illustrate the procedure to them.

V. Summarize lesson.
   A. Review the mathematical terms and concepts covered.
   B. Emphasize paper folding as both mathematical and artistic creations.

VI. Pass out assignment.

ANSWERS for assignment can be found in the background material.

8) a pentagram is formed

ASSIGNMENT

1) Take a square sheet of paper and figure out a way to fold it in order to form an isosceles triangle (a triangle with two congruent sides) from the creases.

2) From a square sheet of paper, determine a way to fold it in order to form an equilateral triangle (all three sides congruent) from the creases.

3) From a square sheet of paper, determine a way to fold it so its creases form a regular hexagon (a 6-sided polygon with all sides congruent).

4) Take a rectangular sheet of paper and follow the diagram. You will form an origami paper popper.

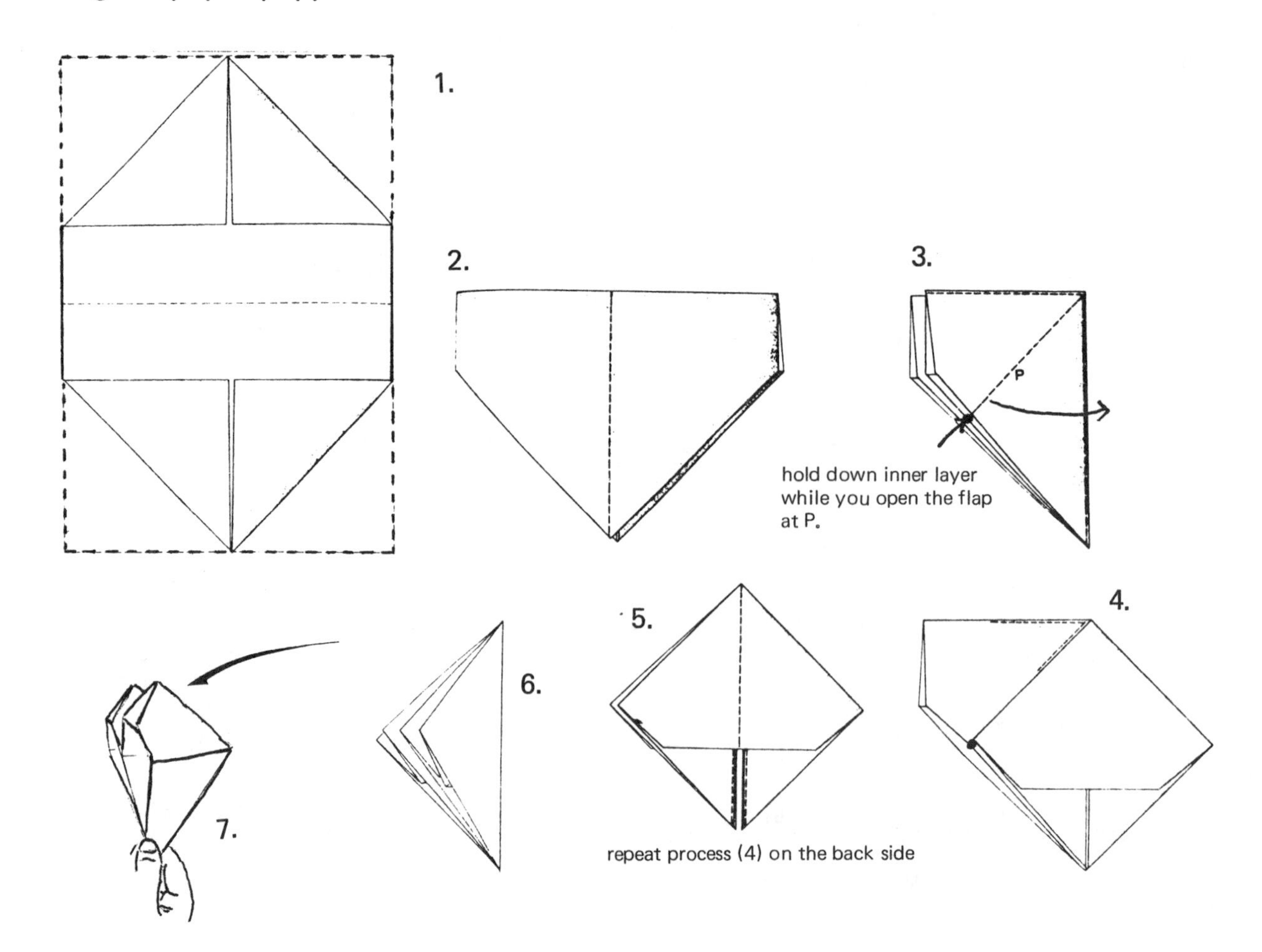

5) Take a sheet of paper, and divide each of its angles into four congruent angles. Using the creases, determine a way to form an octagon (an 8-sided polygon).

6) Using a square sheet of paper, follow the diagram. You will make an origami lotus flower.

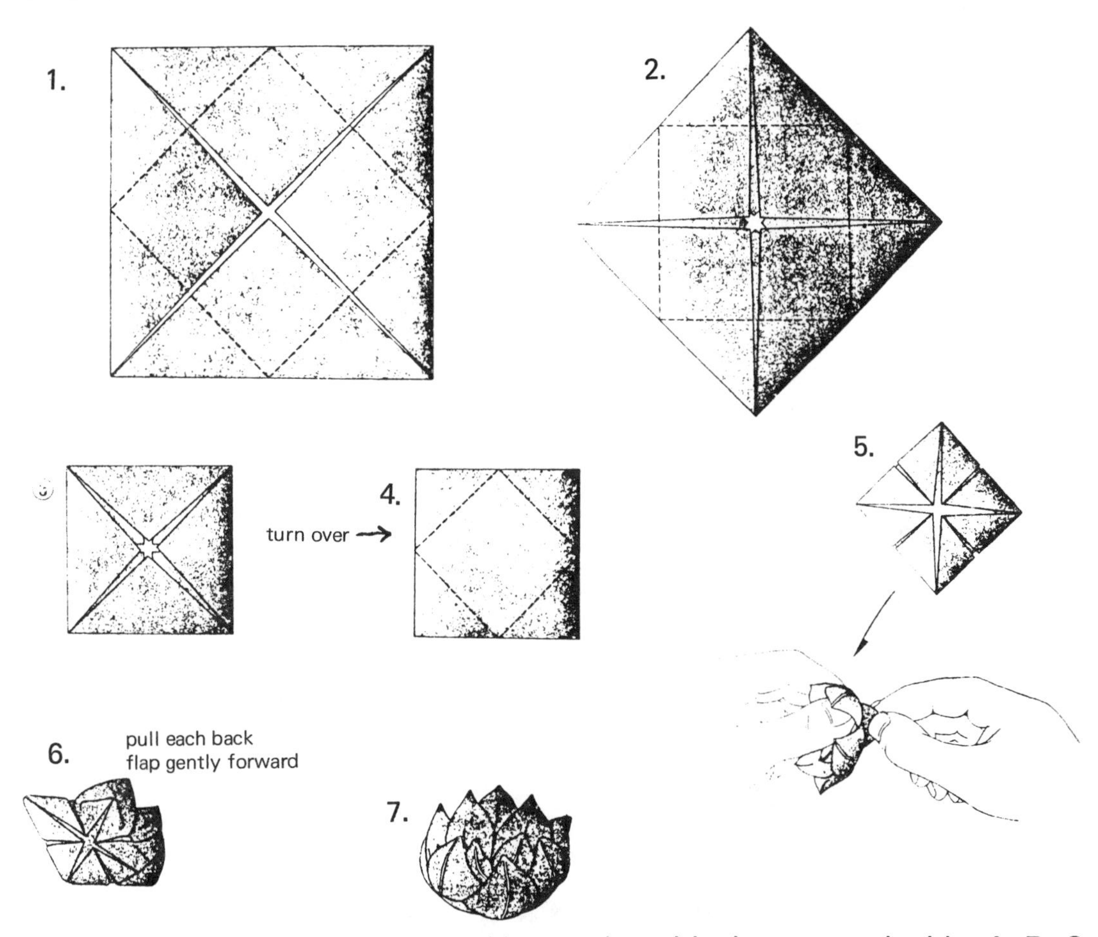

7) Take a square sheet of paper and locate the midpoints on each side, A, B, C, D. Make the crease as indicated in the diagram. Cut along the creases. Now fit the pieces together to form five squares.

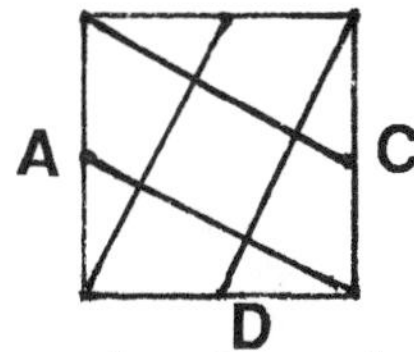

8) From a long strip of paper, form a knot as pictured. Now hold this paper up to the light. What is formed?

NON-EUCLIDEAN GEOMETRIES

an abstract design of Poincaré's hyperbolic geometric model

non-Euclidean geometries

BACKGROUND MATERIAL

The 19th century was a period of revolutionary ideas in politics, art, science, and mathematics as well. The discovery and development of non-Euclidean geometries marked the beginning of modern mathematics in the same way impressionistic painting marked the beginning of modern art.

EVOLUTION OF EUCLIDEAN GEOMETRY

Thales (640-546 B.C.) was considered the first to take a logical approach to geometric ideas. Others followed over a period of about 300 years, discovering much of the geometry studied in high school. In about 300 B.C., Euclid collected and organized all the geometric ideas that had been created. It was an enormous task. He compiled all this information into a mathematical system which has come to be known as *Euclidean geometry*. In his book, **The Elements**, he arranged the information so that it followed a logical progression. **The Elements**, written more than two thousand years ago, is far from a perfect mathematical system when scrutinized by today's mathematicians; but it remains a phenomenal piece of work.

EVOLUTION OF NON-EUCLIDEAN GEOMETRIES

Euclid's 5th postulate, known as the **Parallel Postulate**, was not readily accepted by many mathematicians as a postulate. Instead, many felt it could be proven from previous theorems and postulates, and thus was not a postulate. For 2000 years mathematicians tried to prove the Parallel postulate by using the first four postulates and their subsequent twenty-eight theorems from Euclidean geometry.

In the 19th century certain mathematicians tried to prove the Parallel Postulate by various indirect proofs. To do this, the opposite of the Parallel Postulate was assumed, and they proceeded to reason logically trying to reach a contradiction. Instead of reaching a contradiction, they stumbled upon new mathematical systems– new forms of geometries– competely independent, consistent, logical, yet so very different from Euclidean geometry. These came to be known as **non-Euclidean geometries.**
The best known of these are:

1) ***hyperbolic geometry*** by Lobachevsky and Bolyai
2) ***spherical (and elliptic) geometry*** by Riemann

DISCOVERIES OF NON-EUCLIDEAN GEOMETRIES

In today's language, *The Parallel Postulate* states that–

> *Through any point not on a given line, there is only one line parallel to the given line.*

The negations of the Parallel Postulate that were used in an attempt to prove it by an indirect method were:

> ***I.*** *Through any point in a plane, there are at least two lines parallel to the given line.*
>
> ***II.*** *Through any point there are no lines which can be drawn parallel to a given line.*

The first unpublished account of the discovery of a non-Euclidean geometry was done by the Italian Jesuit, Saccheri in 1733. He made an assumption similar to ***I.*** with the intention of proving it false to establish the Parallel Postulate as the only alternative. But he did not succeed in finding any contradictions to his assumptions from any of the propositions that followed from ***I.*** Instead, he discovered ideas so strange and alien to Euclidean geometry and the physical world as it was perceived, that he did not accept or disclose his findings.

Carl F. Gauss (1777-1855) also tried to prove the Parallel Postulate using ***I.***, and realized in the process that he had discovered a new geometry. He feared the ridicule and controversy of making his revolutionary ideas public. So it was not until Nicolai Lobachevsky (1793-1856), a Russian mathematician, and Johann Bolyai (1802-1860), a Hungarian mathematician, that proposals for the new geometry were published. They arrived at their surprising conclusions independently–Lobachevsky in 1829 and Bolyai in 1832, both working with ***I.*** The first type of non-Euclidean geometry came to be known as Lobachevskian-Bolyain geometry or **hyperbolic geometry.** This geometry can best be illustrated by the following diagram.

Given line L and point P not on L, there exist at two lines through P parallel to L.

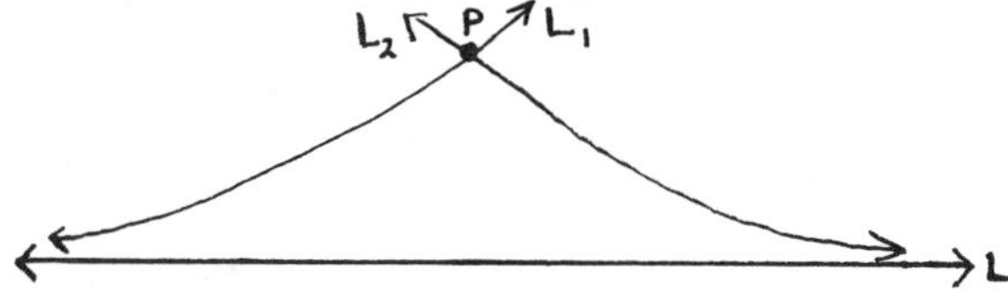

The lines, L_1 and L_2, are not parallel in the same way we expect lines to be parallel in Euclidean geometry. They do not remain equidistant throughout.

The distance between *hyperbolic parallel lines* diminishes in one direction and increases in the other direction. Thus *hyperbolic parallel lines* are *asymptotes* (curves that continually approach in a line but never touch or cross it).

Euclid's geometry led to the geometry of the plane. Hyperbolic geometry leads to the geometry of the pseudosphere– a curve rotated about a line to which it is asymptotic.

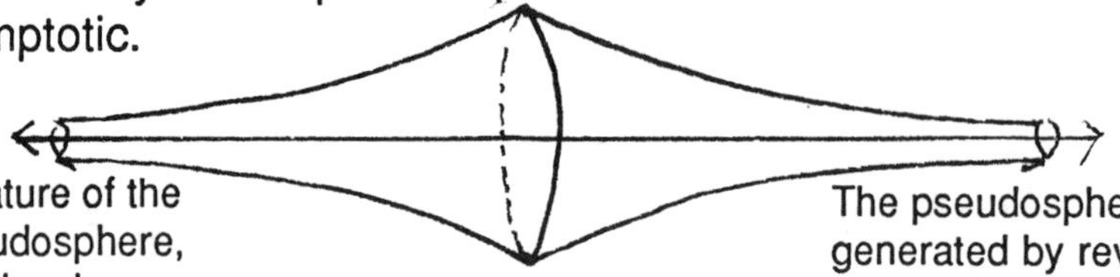

Because of the nature of the shape of the pseudosphere, lines on the pseudosphere are actually curves.

The pseudosphere surface is generated by revolving the curve known as a **tractrix** about a horizontal line.

Now consider what kind of geometry evolves when statement ***II.*** is used in an indirect proof. This was done by G.F. Bernhard Riemann in 1854. Other mathematicians who considered ***II.*** found contradictions of Euclid's other assumptions. But Riemann decided to see what would happen if he also discarded another postulate of Euclid, i.e. *a straight line may by produced to any length in a straight line,* or rather he gave it another interpretation. *In Riemann's non-Euclidean geometry a line is boundless but not infinite,* i.e. it has no ends but can be wound around itself and thus is finite in length. Such a geometry can exist on a sphere and so *all lines are great circles which intersect in two points and no lines (great circles) are parallel.*

COMPARISON OF SOME PROPERTIES FROM THE GEOMETRIES

EUCLIDEAN GEOMETRY –Euclid 300 B.C.	**HYPERBOLIC GEOMETRY** –Lobachevsky-Bolyai 1829	**SPHERICAL/ELLPTIC GEOMETRY** –Riemann, 1854
Given a point not on a line, there is one and only one line through the point parallel to the given line.	Given a point on a given line, there are infinitely many (asymptotic) lines through the point that are parallel to the given line.	There are no parallel lines.
Parallel lines are everywhere equidistant.	Parallel lines never intersect but approach each other asymptotically, i.e. distance between them continually diminishes as they are extended, but they never touch.	Every pair of lines intersect.
The sum of the angles of a triangle totals 180 degrees.	The sum of the angles of a triangle totals less than 180 degrees.	The sum of the angles of a triangle totals more than 180 degrees.
As a triangle's area increases, the sum of the angles remains constant 180 degrees.	As a triangle's area increases, the sum of its three angles decreases.	As a triangle's area increase, the sum of its angles increases.

LESSON

INTRODUCTION

Have your students step into a world that can be considered fantasy or reality. Have them look at the geometric nature of our world from a different point of view in the following manner.

Draw a curve on the board.
Ask them what they see. Is it a curve? Propose to them: *If you were a bug and you landed on this curve, could you tell if it were curved from being that close to it?* **So perhaps in reality we are so close to different objects in our world that we do not really see them as they actually exist in the universe.**

Point out to them that the objects studied in high school Euclidean geometry do not really exist in our world. All the objects in geometry have only representations in our world. A square can only be roughly drawn, since it is physically impossible to draw an exact square. A dot only represents a point, since a point has zero dimension and cannot be seen. All the theorems, postulates, definitions in geometry serve as guide lines in our world. They can merely represent what exists or takes place.

Now discuss the historical evolution of both Euclidean and non-Euclidean geometries by referring to the background material.

Highlights from the background material:

Euclidean geometry goes back to ancient Greece when mathematicians, philosophers, and scientists were discovering geometric ideas and proving them. Euclid assembled the geometric ideas and arranged them in a logical order (including undefined terms, definitions, theorems, postulates) so that ideas were deduced logically from previously proven ones. Although high school geometry books are not a direct translation of Euclid's **Elements,** they are based on what he compiled over 2000 years ago. But there was one part of Euclid's geometry that puzzled and intrigued mathematicians over the centuries–***The Parallel Postulate*** (see background material). Many mathematicians doubted whether Euclid was right in assuming it was a postulate. As a result, they tried to prove it (see background material). In their attempts to prove it, they discovered some strange types of mathematics.

These geometries, at first glance, seemed bizarre and impractical, but when one looks at the universe in view of these geometries, one begins to wonder whether Euclid's geometry is realistic. These new geometric systems came to be known as non-Euclidean geometries. Referring to the background material for details, present hyperbolic and spherical geometries.

Point out that the mathematicians that discovered them did so by trying to prove the *Parallel Postulate* by indirect proofs. In general terms, explain the process of indirect proof to your class, gearing your explanation to your class' level of mathematics.

Highlight the following ideas:

In hyperbolic geometry, two lines can be drawn through a point parallel to a given line, not containing the point. The lines in hyperbolic geometry are not lines as we know them–they are actually curves.

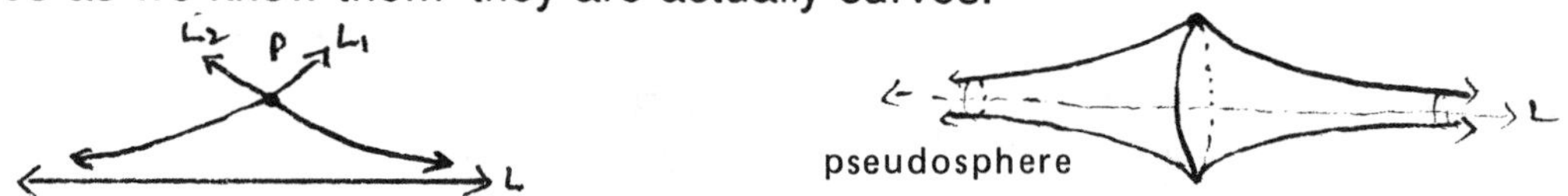

Parallel lines on one side continually get closer and closer to each other, but never intersect (they are asymptotic). The drawing of the pseudosphere above is an example of a three dimensional object containing such lines. As a result, lines bend or are curved, so the shortest distance between two points is not a straight line, but a curved line.

Ask your students to imagine in what kind of universe such things exist.

After considering various student theories, discuss Henri Poincaré's model. Poincaré (1854-1912) was a French mathematician who described a world in which hyperbolic geometry could exist.

POINCARE'S MODEL OF HYPERBOLIC GEOMETRY

Draw a circle on the board and explain this world exists in the interior of this circle. The hottest temperature is located at the center. As one walks toward the boundary of the circle, the temperature decreases and reaches absolute zero at the boundary.

Ask: ***Who lives here?***

The inhabitants are not sensitive to heat, and are not aware of temperature changes. Assume the size of an object changes as it moves in the universe. Every object and creature expands and contracts in proportion to a change in temperature.

Ask: ***Do the inhabitants notice the size change?***

Since every object is changing, no one realizes it is happening. Things shrink as they approach the boundary, and enlarge as they approach the center.

Ask: ***Can an object or inhabitant ever reach the boundary of this universe?***

As a creature moves toward the boundary, its size decreases. The size of its steps become shorter. The nearer it comes to the boundary the more it shrinks. Thus, it will never reach the boundary.

Ask: ***Does the boundary ever appear closer to the creature?***

Since the inhabitants do not realize that the size of their steps and every object are changing as they move, the boundary never appears to get closer but remains infinitely distant..

Ask: ***What is the shortest distance between points A and B.***

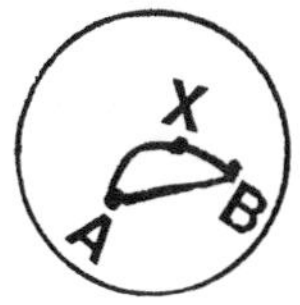

The arc $\overset{\frown}{AXB}$ would be the shortest distance in this universe, since the inhabitants would feel they are approaching B faster along $\overset{\frown}{AXB}$ rather than $\overline{AB}$. The size of their steps is larger along $\overset{\frown}{AXB}$ because they are closer to the center, where the temperature is greater.

An interesting analogy you may want to present is between the geometric model of Poincaré and Einstein's theory that objects in the universe shrink as they approach the speed of light.

Now consider the non-Euclidean geometry of Riemann, which also developed from a indirect proof of the Parallel Postulate–see backgound material.

Have the students reconsider their view of the term *line*. Consider a universe that exists on a sphere. Draw a sphere on the board in order to demonstrate the properties of this geometry. In this universe there are no straight lines, since a line is defined as any great circle of the sphere. A circle has no beginning and no ending, it is boundless–one can never reach its end. Yet it is finite, since its length is its circumference.

Have a student draw two great circles on the sphere. Have the students discover that *any two great circles must always intersect–in two points.* Thus, they can conclude that there are no lines (great circles) in this universe that are parallel !

Refer to the comparison chart from the background material for additional properties which would be appropriate for your class.

In your conclusion, point out that the evolution of Euclidean and non-Euclidean geometries are beautiful examples of how mathematics evolves, expands and is applied.

Point out that non-Euclidean geometries have begun to find their place in the universe. Their properties and concepts may prove better able to describe universal phenomena than Euclidean geometries.

LESSON'S OBJECTIVES

1) Student will learn the historical evolution of Euclidean geometry.

2) Student will learn the historical evolution on non-Euclidean geometries.

3) Student will learn about the discoveries of hyperbolic geometry and spherical geometry.

4) Student will compare these non-Euclidean geometries to Euclidean geometry.

5) Student will learn of model universes in which these non-Euclidean geometries exist.

6) Student will learn that the objects of Euclidean geometry are no more real than those of non-Euclidean geometry.

7) Student will create his or her own model universe in which the properties of non-Euclidean geometries exist.

LESSON'S OUTLINE

I. Introduction
  A. curve on board
    1) bug's view point

  B. Euclidean objects
    1) do not exist in our world
    2) Euclid and **The Elements**

II. Euclidean geometry
  A. Evolution
    1) Thales
    2) Euclid
      i. **The Elements**–organization of a mathematical system – geometry
  B. Parallel Postulate
    1) Statement I. and II. from background material
    2) indirect proofs of the Parallel Postulate
      i. using statement I. leads to hyperbolic geometry
      ii. using statement II. leads to spherical geometry

IV. Hyperbolic geometry
  A. evolution from statement I.
    1) Lobachevsky and Bolyai
  B. properies
    1) lines are curves–asymptotes
    2) two lines can be parallel to given line through a given point off that line
    3) other properties from background material–teacher's discretion

  C. Poincaré's hyperbolic model
    1) circle–temperature, size
    2) What happens to inhabitants and objects in this universe?
      i. boundary
      ii. distance between two points
    3) Einstein's theory with relation to Poincaré's model

V. Spherical geometry–Riemann
  A. evolution from statement II.
    1) Riemann

B. properties

1) lines are great circles

i. intersections

ii. boundless

iii. finite

2) no lines are parallel in this geometry

3) other properties from background material–teacher's discretion

VI. Summary

VII. Pass out assignment.

ANSWERS to assignment on non_eulcidean geometries

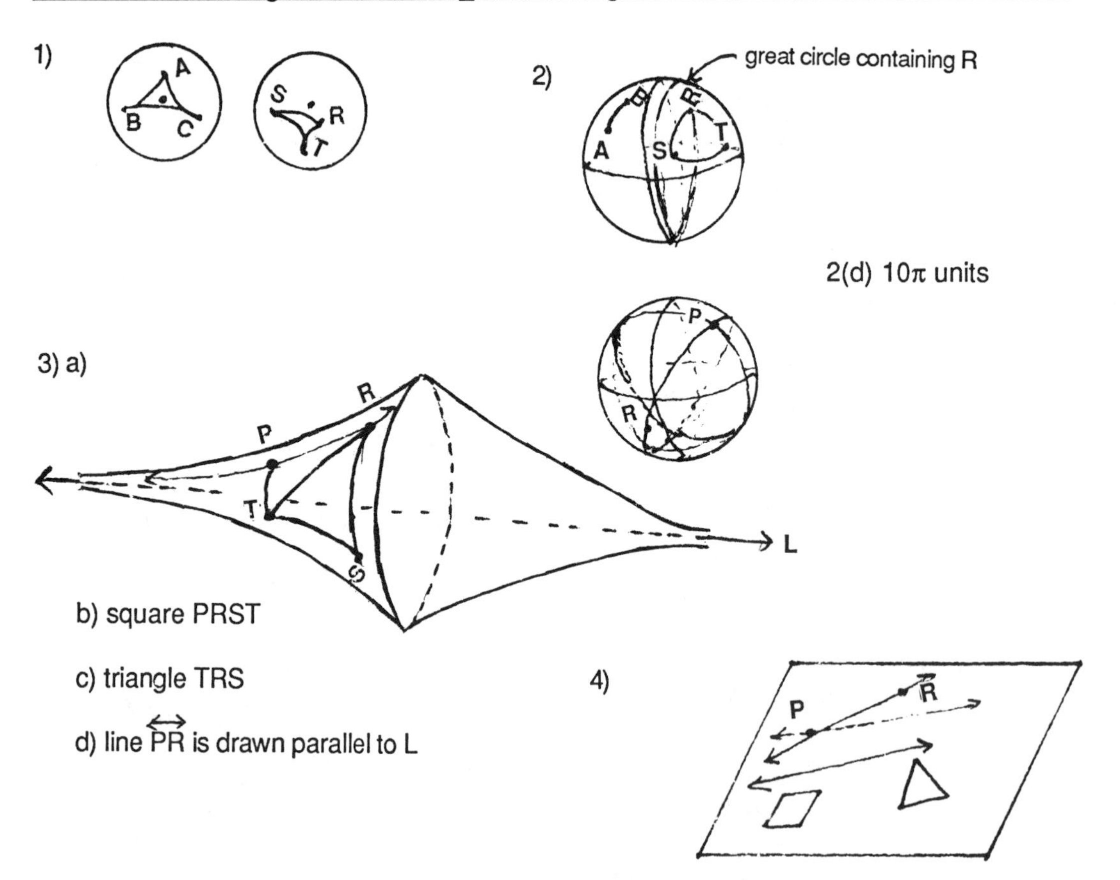

ASSIGNMENT

1) Assume the circle below is Poincaré's hyperbolic geometric model universe.

a) Draw in $\triangle ABC$ as it should exist under the conditions of this model universe.

b) Draw in RST as it should exist under the conditions of this model universe.

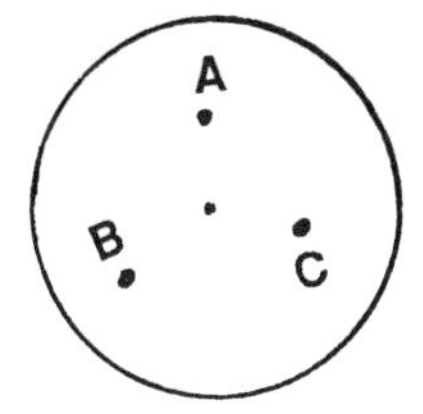

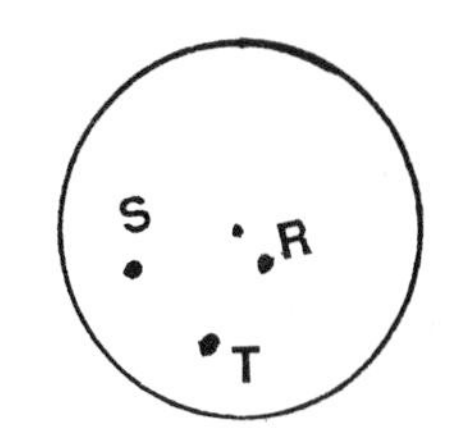

2)Referring to Riemann's spherical non-Euclidean geometry, draw the following objects as they should exist:

a) segment AB

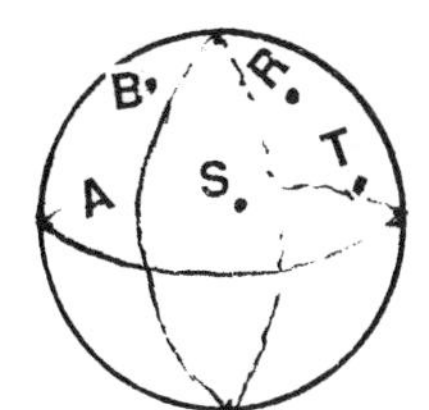

b) $\triangle RST$

c) a line passing through R

d) If the radius of the sphere is 5 units, what is the length of line R?

e) Can you draw a line passing through P and intersecting line PR? If so, do it.

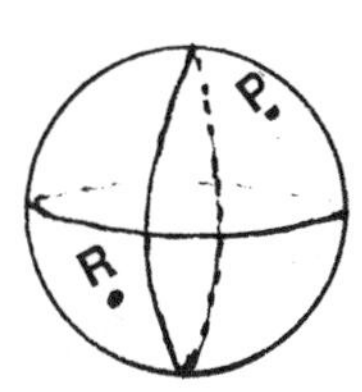

3) The drawing below is called a pseudosphere. The objects and properties of hyperbolic geometry can exist on this surface. Draw in the following objects. Be sure to conform to the shape of the pseudosphere.

a) line PR

b) draw a square as it would appear on the pseudosphere

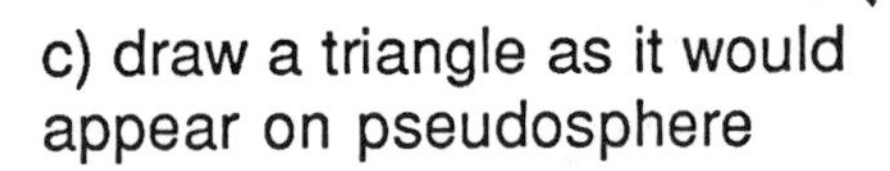

c) draw a triangle as it would appear on pseudosphere

d) draw a line parallel to L and passing through P

4) The drawing below is a plane. The objects and properties of Euclidean geometry can exist on a plane. Draw in the following objects. Be sure to conform to the shape of the plane.

a) line PR

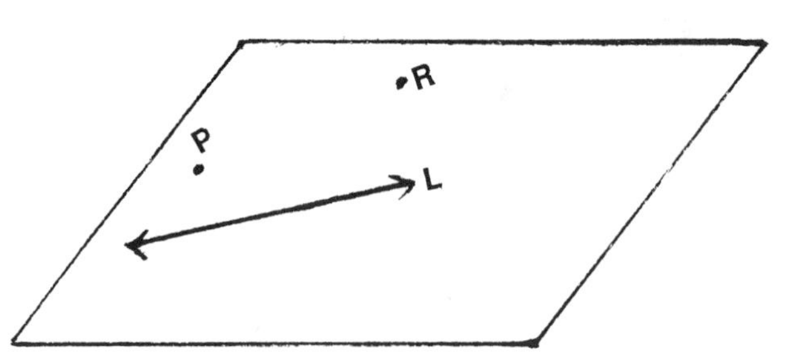

b) draw a square at it would appear on the plane

c) draw a triangle as it would appear on the plane

d) draw a line parallel to L and passing through P

e) Can you draw another line parallel to L and passing through P? If so do it.

OPTICAL ILLUSIONS

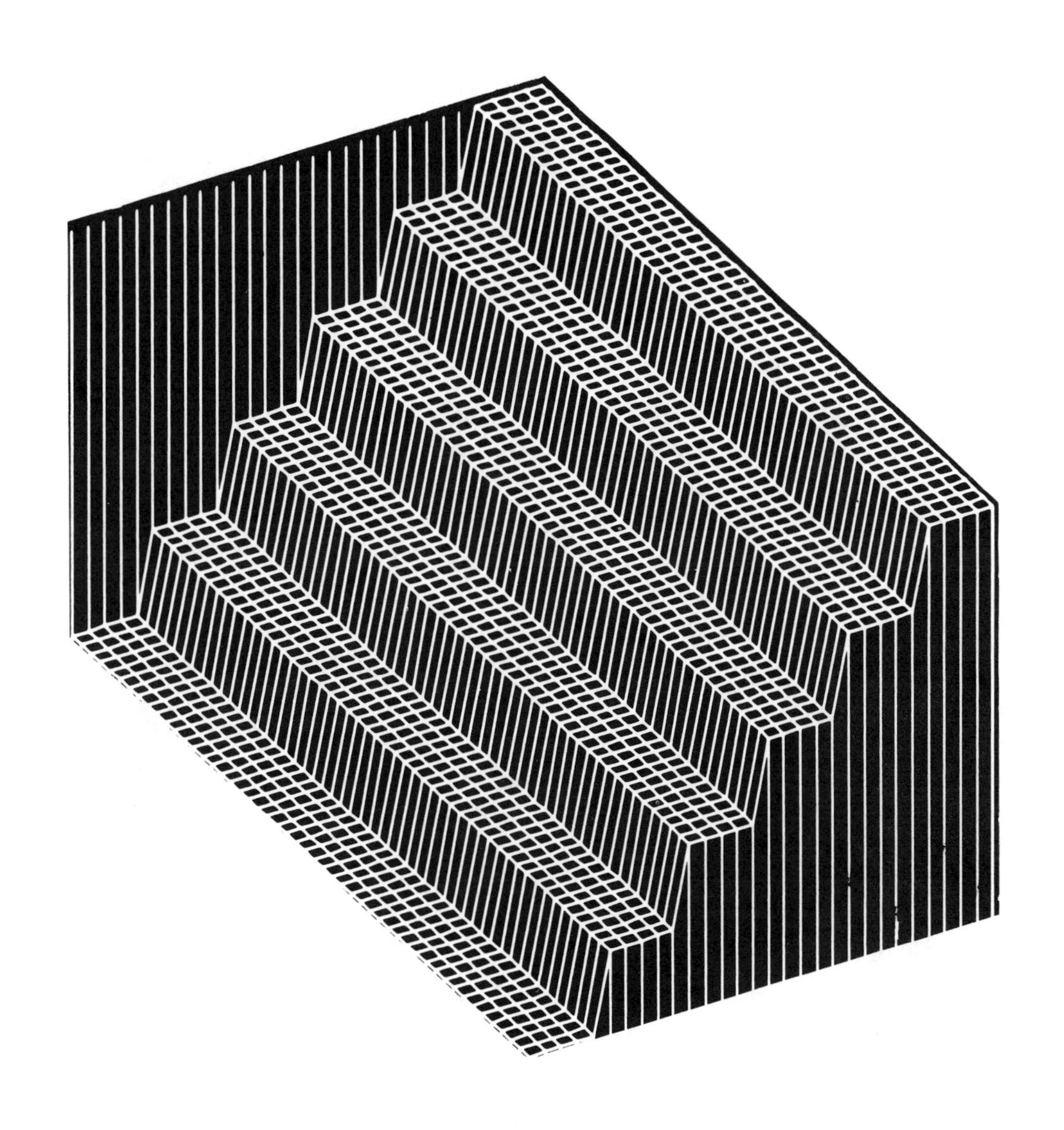

OPTICAL ILLUSIONS

BACKGROUND MATERIAL

Johann Zollner (1834-1882) stumbled upon a piece of fabric with a design similar to the optical illusion below:

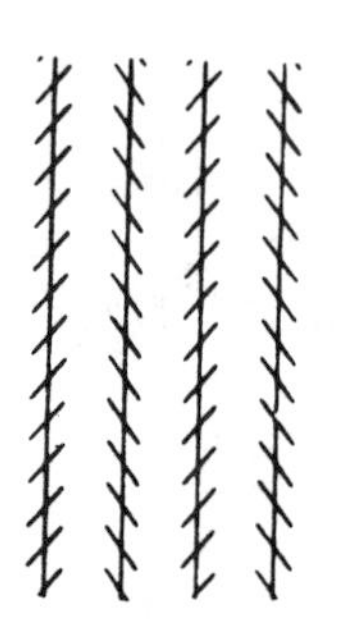

As a result, this illusion triggered the study of optical illusions during the second half of the 19th century. At this time there was such a surge of interest in the field of optical illusions and why they occur.

A brief list of the scientists who studied optical illusions with some their contributions in their fields of expertise:

HERMAN L. F. von HELMHOTZ (1821-1894) was a physiologist, physicist, and mathematician. He was fortunate to study under Johannes Muller and received a degree in neurophysiology.

Helmhotz was the first to measure the speed of nerve impulses, and published **Physiological Optics,** a very important study on the physics and physiology of vision. He also laid the foundation for the science of acoustics with his publication **On the Sensation of Tone.** Helmhotz was the inventor of the opthalmoscope (1850) and the opthalmometer (1864), used to measure eye muscle activity. His outstanding work in electricity and mathematics aided in the formation of the ether vortex theory of the atom. His exceptional work in non-Euclidean geometry opened the way for the theory of relativity.

EWALD HERING (1834-1918) was a physicist and psychologist who studied nerve function and color perception.

JOHANNES PETER MULLER (1801-1858) was a physician and one of the founders of experimental physiology. He was very inspirational to his students, and the first pathologist to use the microscope to examine diseased tissue.

ALBERT OPPEL (1831-1865) was a geologist, paleontologist, and one of the earliest stratigraphers. He discovered that paleontologic and lithologic zones do not have to be identical or dependent.

WHILHELM WUNDT (1832-1920) is regarded as the father of scientific psychology. Besides being a professor of philosophy, he opened the first experimental psychology laboratory in the world. His main thrust was to seek scientific knowledge through observation of social behavior, especially of customs and language systems of ethnic groups.

JOHANN K. F. ZOLLNER (1834-1882) was an astrophysicist. As a professor of astronomy he made many contributions to the study of comets, the constitution of the sun, and the thermal conditions of planets.

He invented the photometer (or artificial star), which made possible the visual measurement of the size of stars. Later in his life he worked in metaphysics, spiritualism, and hypnotism.

The following is a list of the optical illusions discussed in the lesson with a description and example of each illusion, and a plausible explanation for its occurence.

FULL vs EMPTY SPACE ILLUSION

Identical space or distance occupied by objects appear unequal. This illusion is created when the space or distance occupied by identical objects is filled in or left blank.

CONVERGENCE/DIVERGENCE ILLUSION

Objects of the same size appear unequal. This illusion is created by angles or segments which lead our eyes inward or outward, and thus shorten or lengthen an object.

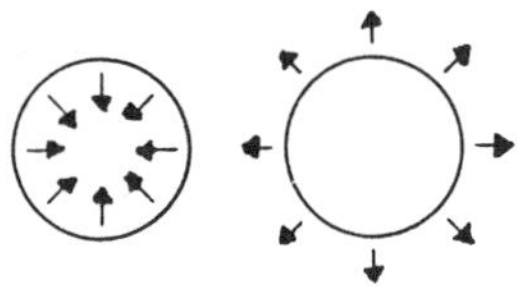

BISECTION ILLUSION

This illusion is created by the location of a vertical object over a horizontal object. The horizontal object is made to appear shorter.

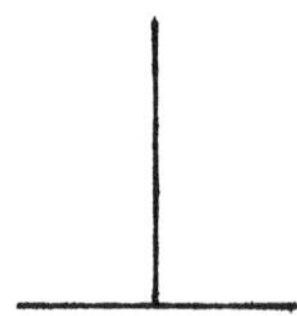

IRRADIATION ILLUSION

This illusion is created when light and dark regions are together. Even if the regions are identical, the light region will appear larger. This is due to the eye's structure. The image of the light region on the retina radiates into the dark region; and thus makes the dark region appear smaller.

PERSPECTIVE ILLUSION

When identical objects are placed in different locations in a perspective drawing, the objects are made to appear different in size.

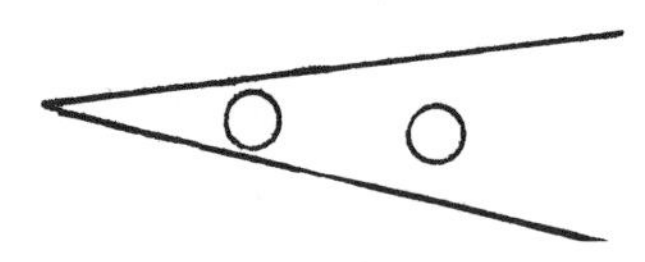

OSCILLATION ILLUSION

This is an illusion which makes our eyes shift back and forth between two or more interpretations of a diagram.

POGGENDORFF or CROSS-BAR ILLUSION

This type of illusion employs the use of a vertical band and diagonal broken segments. The illusion is in determining which segments aline with which.

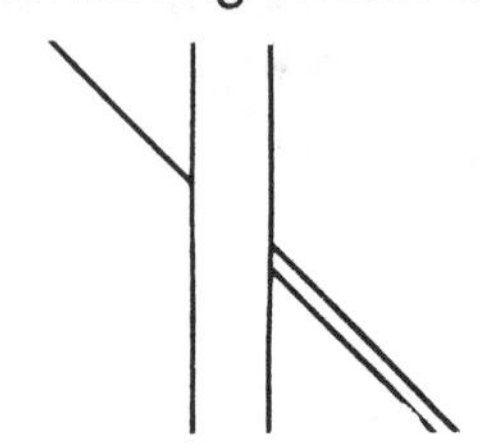

ZOLLNER'S ILLUSION
This illusion makes parallel lines appear non-parallel by using diagonal segments to intersect the parallel lines at different angles.

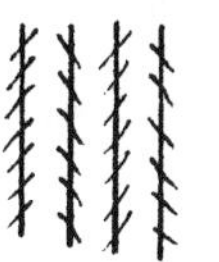

Some of the areas in which optical illusions appear are:

1) nature–Some of our first encounters with optical illusions may be those created by nature, such as the use of camouflage.

2) architecture–Architecture is another area where we come across many optical illusions. Architects from ancient times found that structures built precisely straight did not end up appearing straight to our eyes. Because of the physical structure of our eyes, for example the curvature of the retina, straight lines falling on each other at particular angles may appear to curve when our eyes view them. One famous example is the Parthenon. The architects during this time realized the optical illusion that would have been created had straight columns been used on the Parthenon. As a result, the columns of the Parthenon actually curve outward, as do the sides on the rectangular base of the Parthenon. Thus, by making these compensations, the structure and the columns appear straight and aesthetically pleasing.

3) fashion– Optical illusions also play an important part in the fashion world. By certain choices of clothing- the lines, the cut, the colors- we can flatter our body by changing how others perceive it. For example, the placement of pockets on trousers can alter the perception of the size of the buttocks. The design on the waist of a garment influences how thick the waist appears. A black dress makes one appear more slender than a white dress of the same design.

4) marketing–Many people are not aware of the use of optical illusions in the packaging and marketing of products. Some containers' shapes may even be purposely inverted or changed in some other way to make it difficult for the buyer to make visual comparisons.

LESSON

preparations: overhead projector and transparencies of enclosed optical illusions ***or*** your own drawing of these illusions.

A fascinating way to introduce this lesson is to cut out the two congruent shapes provided below. Holding one above the other on the chalk board or on the overhead projector, ask your class which appears larger. Their response will undoubtedly be the lower one. After they have given their reply, slide the lower one above the upper one and repeat the question. I suggest you do not discuss the causes of this illusion at this point. But for your reference later on, it was due to **convergence/ divergence** –discussed in the background material. Our eyes are guided (converge) to the top of the lower pie by the position and slant of the pies.

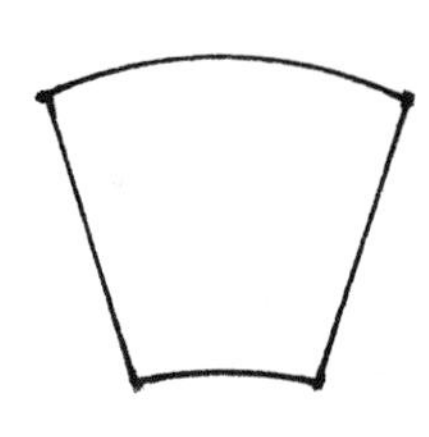

What we see is not always what exists. And it is especially important in studying mathematics and science not to base our conclusions strictly on what we perceive, but rather to verify by actual measurement.

We often come across optical illusions in our everyday lives. Sometimes we are able to detect them, but more often we are unconscious of them. We will be seeing some exciting and baffling optical illusions, and we will try to explain why they occur.

First see if students can come up with some examples of optical illusions in the following areas:

1) nature
2) architecture
3) fashion
4) marketing

If students have trouble coming up with some examples, refer to the background material.

This is a good point at which to introduce some *historical information*. It was in the 1800's that a keen interest arose in the field of optical illusions. It was especially the physicists and the psychologists who devoted much time and effort to discover and explain various optical illusions.

At this point it might be a good idea to put a list of the optical illusions on the board in the order they will be covered. This will be especially helpful to students since most of these names will be alien to them.

full-empty space
convergence/divergence
bisection
irradiation
perspective
oscillation
Poggendroft or cross-bar
Zollner's

The most enjoyable way to present the various illusions is to show them, give the students some silent time to individually study the diagram, and then ask for the illusions they perceive.

I. optical illusion of full-empty space

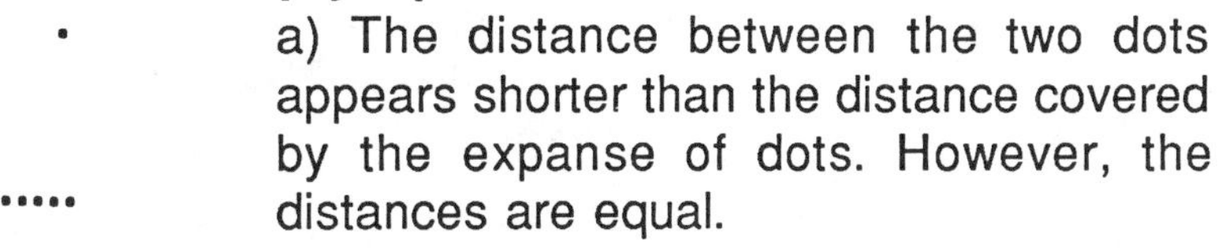

a) The distance between the two dots appears shorter than the distance covered by the expanse of dots. However, the distances are equal.

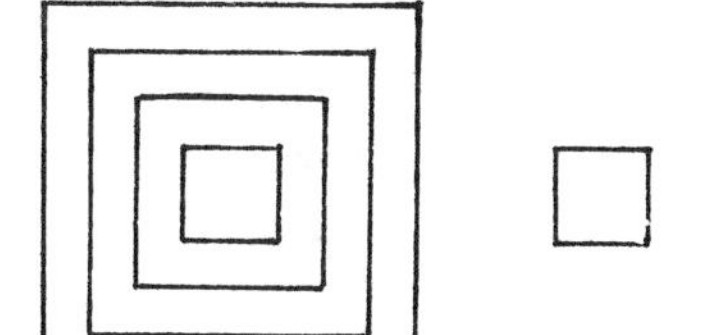

b) The inner empty square appears larger than the same size square filled in with smaller squares.

note: The intensity of the filled in space will affect the illusion.

II. covergence/divergence illusion (Muller-Lyer illusion)

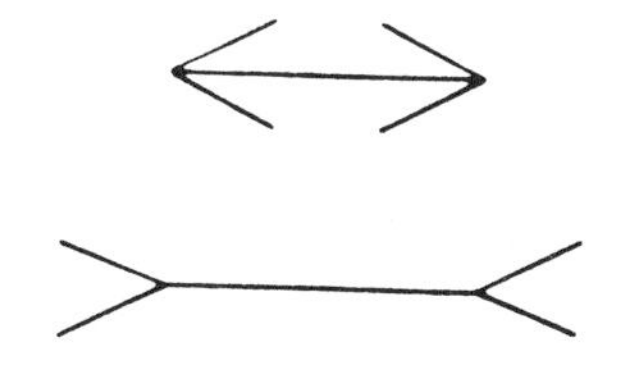

The segments appear unequal. The shorter one appears on top. This illusion was first proposed in 1859 by Zollner and described by Muller and Lyer in 1889. The word convergence here describes how our eyes are led inward by the top segment's arrows. Thus, the top segment appears shorter. Divergence describes the eyes being led outward, so the bottom segment appears lengthened by its arrows.

This illusion is used in fashion design. The placement of pockets or various types of stitching can lead our eyes inward or outward depending on the location of the pockets or the stitching and thus affects the appearance of the size of the person wearing the clothing.

(b) The study of crystals is a good example of where this optical illusion is quite prominent. (**Note:** the same edge of a crystal may appear very different depending upon the angle from where the edge is viewed.) In this particular illusion you can see how important and deceptive the angle at which we view crystals is. Crystals have so many facets that their shapes may seem altered depending upon the position at which they are viewed.

Scientists must be on their toes for illusions of this sort. When making observations, measurements should be taken to verify conclusions, rather than relying on our eyes for estimations.

III. bisection illusion

1) Another striking optical illusion that was identified in the eighteen hundreds was the *bisection illusion* by Oppel and Kundt in 1865.

Which segment appears longer? The vertical segment should appear longer, though both are the same length. There are several factors contributing to this illusion.

a) The horizontal segment is first seen by our eyes because our eyes are lined up horizontally themselves. Thus, they view the vertical segment later and with more effort.

b) The horizontal segment has a tendency to be cut off at the ends by our eyes because the structure of the retina is curved.

c)Finally, the horizontal segment is viewed as two smaller segments since it is bisected by the vertical segment.

2) In this example the bisection of the horizontal line segment is implied by the centering of the vertical structure over the horizontal. (Although the hat's height appears longer than the hat's brim, they are the same length.)

IV. irradiation illusion by Helmholtz

1) Optical illusions are created by our minds, our eyes' structure or a combination of both. For example, when we view a region that has both light and dark objects, the fluids in our eyes are not perfectly clear and light scatters while passing to the retina at the rear of the eye (this is where the eye detects light). As a result bright light or light areas, spill over onto the dark areas of the image that is on the retina. Thus, a light region will appear larger than a dark one of equal size, as in this illusion.

This explains why dark clothing,

especially black, makes you appear more slender than if you were wearing light, or white clothing of the same design. This illusion is called *irradiation* and was discovered by Helmoltz in the 19th century.

V. perspective illusion

Consider the illusion caused by trying to view something in perspective. The mind is influenced by the drawing of perspective lines. Our eyes are being guided on how to look at the object, and as a result some fascinating illusions take place.

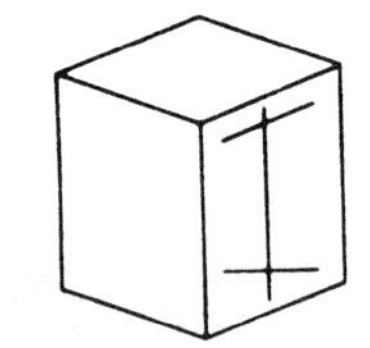

1) In this illusion the actual horizontal and vertical segments which are perpendicular do not appear so. In fact, their angles appear obtuse.

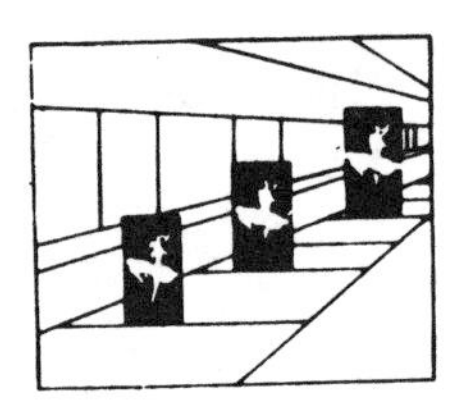

2) Here, the three dancers appear to be different sizes. All three are actually identical, yet they seem to vary depending on their location in the persepctive drawing.

VI. diagonal illusion

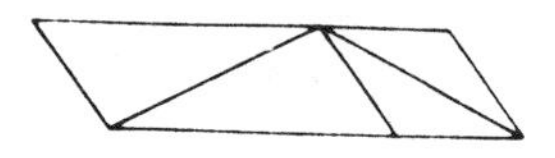

In this illusion it is difficult to pick out what the illusion is. Any guesses?...Look at the diagonals of the parallelograms. They appear unequal in length, yet they are equal.

VII. oscillation illusion

Our minds are also influenced by past experiences and suggestions. The mind at first will see an object one way, and when a certain length of time has passed it will change its point of view. The view varies between individuals. The time factor is influenced by our attention, or how quickly we get bored with what we first zero in on. This kind of illusion is called *oscillation illusion.* In this illusion the mind alternates its attention between different ideas that present themselves in the diagram. As you view the following slides, note when your attention shifts.

a) What did you see first? (Attention roams between the light and dark regions, between the goblet and the profiles.) Each person's mind has a preference as to what it sees first.

b) Here, do you see the tops of the cubes? Was the shaded region a top or a bottom? Are you able to make it become both, but not simultaneously?....This type of oscillating cube illusion is present on the etched surface of a diamond and in aerial photographs. An aerial photo held one way may show a hill, but when turned over the hill appears as a valley.

c) The old/young woman illusion has always intrigued classes. It usually takes awhile to see both figures. Some people get frustrated, but with concentration you should find a young woman and an old woman. (Good luck!) (**Hint:** the old woman's nose is the jaw of the young woman. The necklace of the young woman is the old woman's mouth.)

VIII. cross-bar illusion

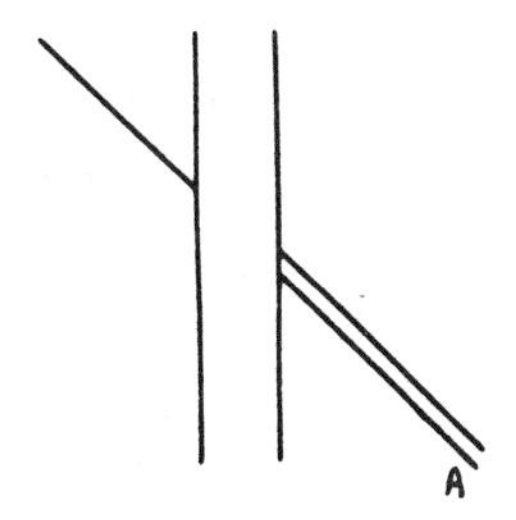

The upper segment alines with which segment below?....(the best way to verify this is with a straight edge. It lines up with A). The factors that contribute to this illusion are:

1) the eyes try to measure the acute

angles as a means of alining the segments;

2) the continuity of the segment is broken by the vertical band.

By experimenting, you can verify that the illusion is intensified if the segments form a more acute angle with the vertical band. The illusion is also diminished as the vertical band becomes thinner.

IX. Zollner's illusion (see background material on Zollner)

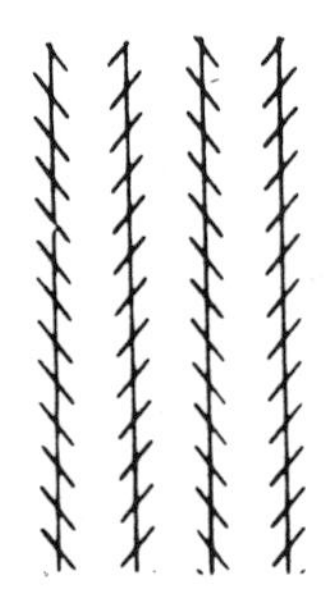

1) This fascinating optical illusion uses parallel lines and acute angles. It was this illusion that triggered the study of optical illusions in the 19th century. The vertical lines are actually parallel, but certainly do not appear so. Some explanations given for this are:

1) the difference between the acute angles which are set in different directions on parallel segments;

2) the curvature of the retina;

3) superimposed segments cause convergence/divergence which make the parallel segments curve.

It was discovered that this illusion is most intense when the diagonal segments form angles of 45 desgrees with the parallel segments.

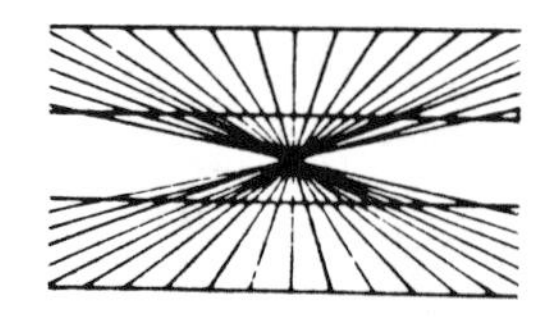

2) Hering's illusion is a modification of Zollner's. When Hering came up with this illusion in 1860, astronomers and physicists were very interested in it because they were afraid it would prove their visual observations unreliable, since

the diagonal segments resembled light rays.

X. miscellaneous illusions

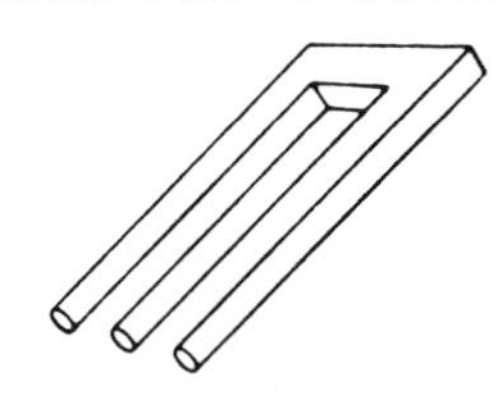

1) ***Two pronged figure*** does not exist in our world; yet because we are viewing it our mind feels we must make sense out of it. Frustrated, our eyes search back and forth for meaning (a type of oscillation illusion).

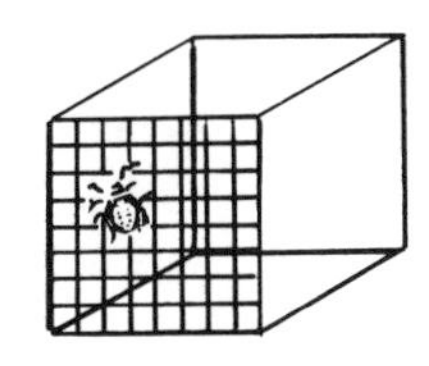

2) ***The insect in the cage illusion*** You'll need to stare at this for some time. Then something strange should happen....Anyone notice yet?....The insect should appear both as being inside and outside the cage. This is another example of oscillation illusion.

Artists often use optical illusions in their works. M. C. Escher is well known for his various uses of illusions. Encourage students to look up his work in the library. I suggest you share with them his work ***Concave and Convex 1955.***

Concluding remarks

It's mind boggling to consider how many distortions pass by our eyes undetected. Yet, it's exciting to realize their existence. The study of optical illusions evolved as a result of individuals' encounters with illusions. It goes back thousands of years. Can you imagine, for example, how the Parthenon would have looked if the architects had not been aware of optical illusions? So the next time you say, "I saw it with my own eyes," or "Seeing is believing," hopefully you will want to verify it by means other than with your eyes.

LESSON'S OUTLINE

I. Introduction

- A. example of convergence/divergence illusion
- B. optical illusions role in our every day life
    - 1) nature
    - 2) architecture
    - 3) fashion
    - 4) marketing

II. Optical illusions, examples, and their causes

- A. full-empty illusion
- B. convergence/divergence illusion
- C. bisection illusion
- D. irradiation illusion
- E. perspective illusion
- F. diagonal or geometric illusion
- G. oscillation illusion
- H. cross-bar illusion
- I. Zollner's illusion
- J. miscellaneous illusions
    - 1) two- pronged–oscillation type illusion
    - 2) insect cage illusion–oscillation illusion

III. Historical background

- A. Zollner's illusion triggered 19th century studies
- B. knowledge of optical illusions in ancient tiimes
    - 1) architects of Parthenon

IV. Pass out assignment.

LESSON'S OBJECTIVES

1) Student will become aware of the role optical illusons play in our every day life–e.g. in nature, architecture, fashion, marketing

2) Student will learn of the different optical illusions and their causes–
full vs. empty space, convergence/divergence, bisection, irradiation, perspective, oscillation, cross-bar, Zollner's illusion

3) Student will be given the historical background of the evolution of the study of optical illusions.

4) Student will become aware that there are distortions caused by various factors, and that it is important not to base our conclusions strictly on what we perceive, but rather to *verify by actual measurement.*

ANSWERS to optical illusion assignment

1)
a) Center circles are of equal size though the one on the right appears larger. This is caused by irradiation.

b) The parallel lines appear to bend inward , and thus appear nonparallel. This is a modification of Zollner's illusion, and was created by Wundt.

c) The two circles are the same size but appear unequal. The illusion was created by Ponzo, and is caused by a perspective drawing- the circle closest to the vertex appears larger.

d) When the lower segment is extended upward it will intersect the segment on the left, but they do not seem alined. This is a form of the cross-bar illusion.

e) Both circles are of equal size, but the one on the right appears larger.
This is caused by the arrows creating convergence/divergence.

f) This is known as Thiery's figure. The figure appears to move in and out. The top appears both as a ceiling and a roof, and the bottom appears as an interior and an exterior floor. It is an oscillation illusion.

g) The vertical line segments are of equal length, but the one closest the vertex appears longer. This is caused by perspective drawing.

h) The sides of the square seem to bend inward. The pattern of concentric circles causes this illusion.

i) The inner angles are actually congruent, but the one on the right appears larger. The illusion is caused because the inner angle on the left is enclosed by larger angles while the inner angle on the right is enclosed by smaller angles.

j) The upper base of the top trapezoid and the upper base of the bottom trapezoid are of equal length, but they appear unequal in length. The illusion is caused by convergence/divergence. The upper figure is lengthened by divergence and the lower figure is shortened by convergence. This illusion and the one used for introducing the slide show were created by Wundt.

2) Drawings will vary.

3) Drawings will vary.

4) Originally the tops and bottoms of the S seem relatively equal in size, but when the paper is turned around the bottom of the S appears much larger.

5) Grey spots appear at the intersection of the white segments. A grey spot does not appear at the intersection on which you are focusing.

6) Answers vary to first question on free hand drawing.

7) The top of the pearls are tangent to an imaginary line, but they seem to curve.

8) The distance from A to B is equal to the distance from B to C. This illusion is a combination of both irradiation and convergence/divergence.

A B C

ASSIGNMENT

1) For each diagram below, determine *what is being distorted.* What do you think is causing the illusion? Whenever possible, you may refer to the illusions discussed in class-
full vs. empty space, convergence/divergence, bisection, irradiation, perspective, oscillation, cross-bar, Zollner's illusion, or your own theory.

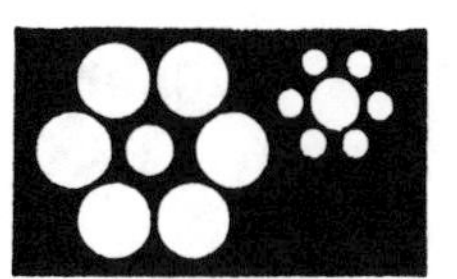

a.

b.

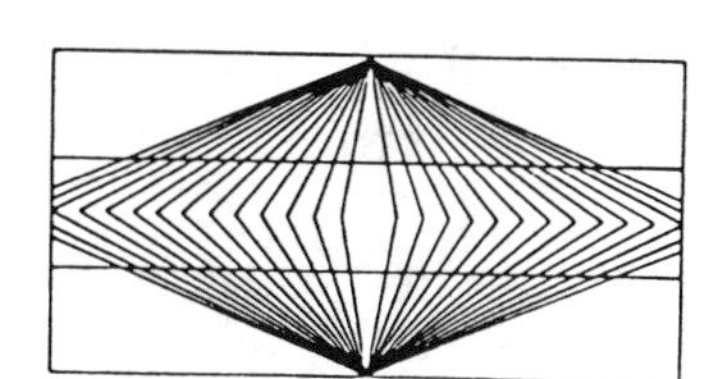

c.

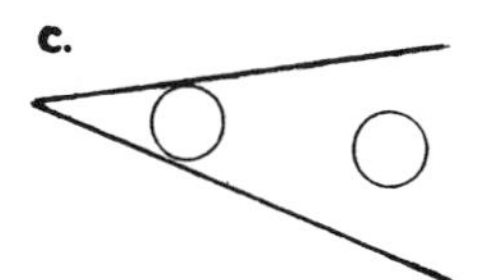

d.

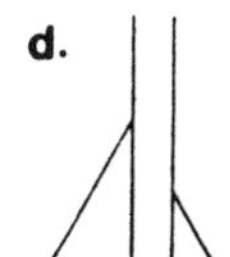

e.

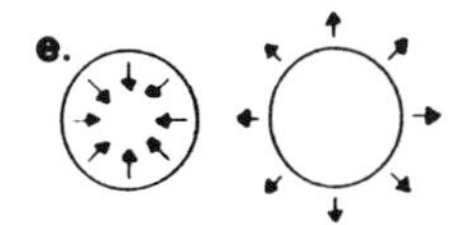

f

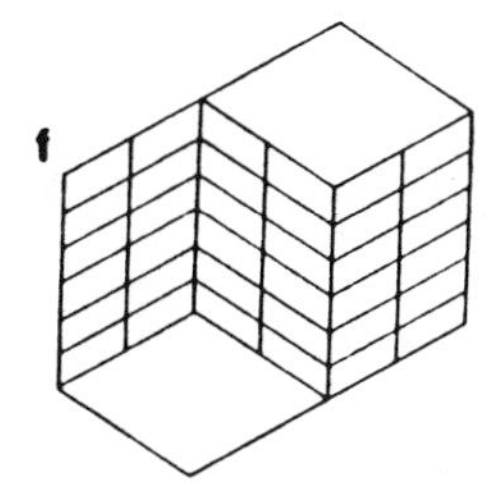

g.

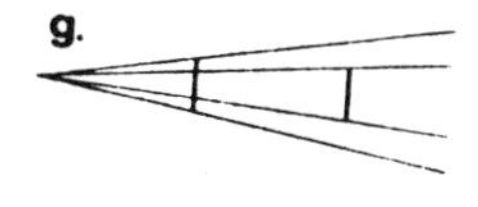

h.

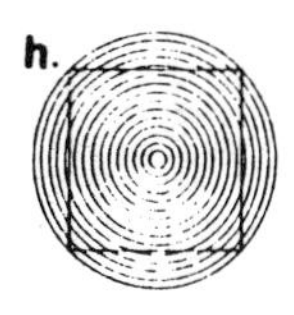

i.

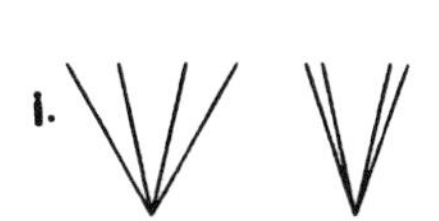

j.

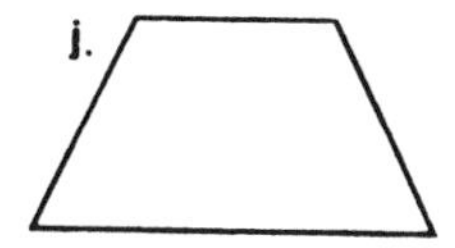

2) From the list of optical illusions in problem (1) draw your own original illusion. It can be a combination of two or more of these illusions.

3)
a) Using a ruler, experiment with the cross-bar illusion. See if you can verify that the illusion becomes more pronounced as the segments form more acute angles with the vertical band. Make at least three drawings.

b) Now draw three examples showing the vertical bands becoming thinner. What happens to the illusion?

4) Our minds are influenced by past experiences. Look at the letters SSSSSSS above. Now turn this paper around so that the letters are upside down. What do you notice?

SSSSSS

5) Stare at the diagram on the left for about 10 seconds. What optical illusion do you notice?

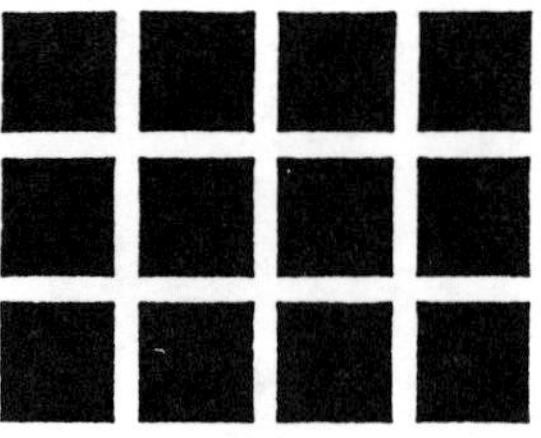

6) By free hand drawing, make a square that appears four times the size of the square below. Now use a ruler and draw a square that is actually four times the size of this one. How does your free hand drawing compare?

7) What optical illusion do you notice in the diagram below? This illusion is called the *Konig necklace.*

8) Look at this for awhile. Let's determine what the illusion is. Measure the length of the horizontal diagonal of the rhombus. Now measure the distance between the two points on the right. What are your measurements?

Could you tell which was larger without measuring?

____

Which illusions from the list in problem (1) are illustrated by this example?

____

____

1a

1b

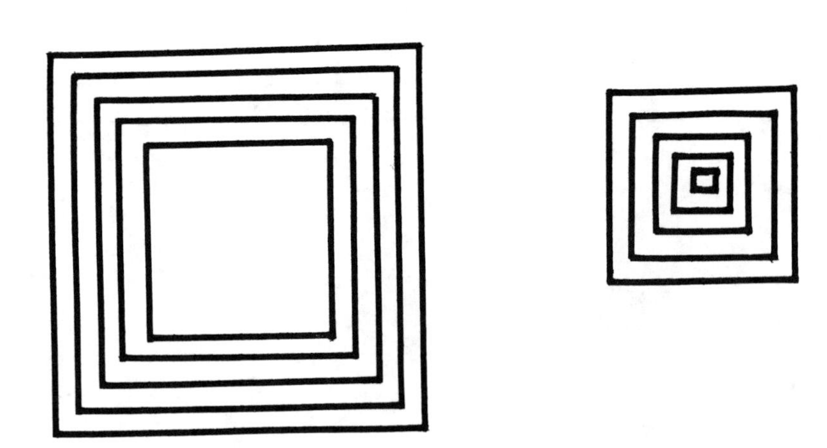

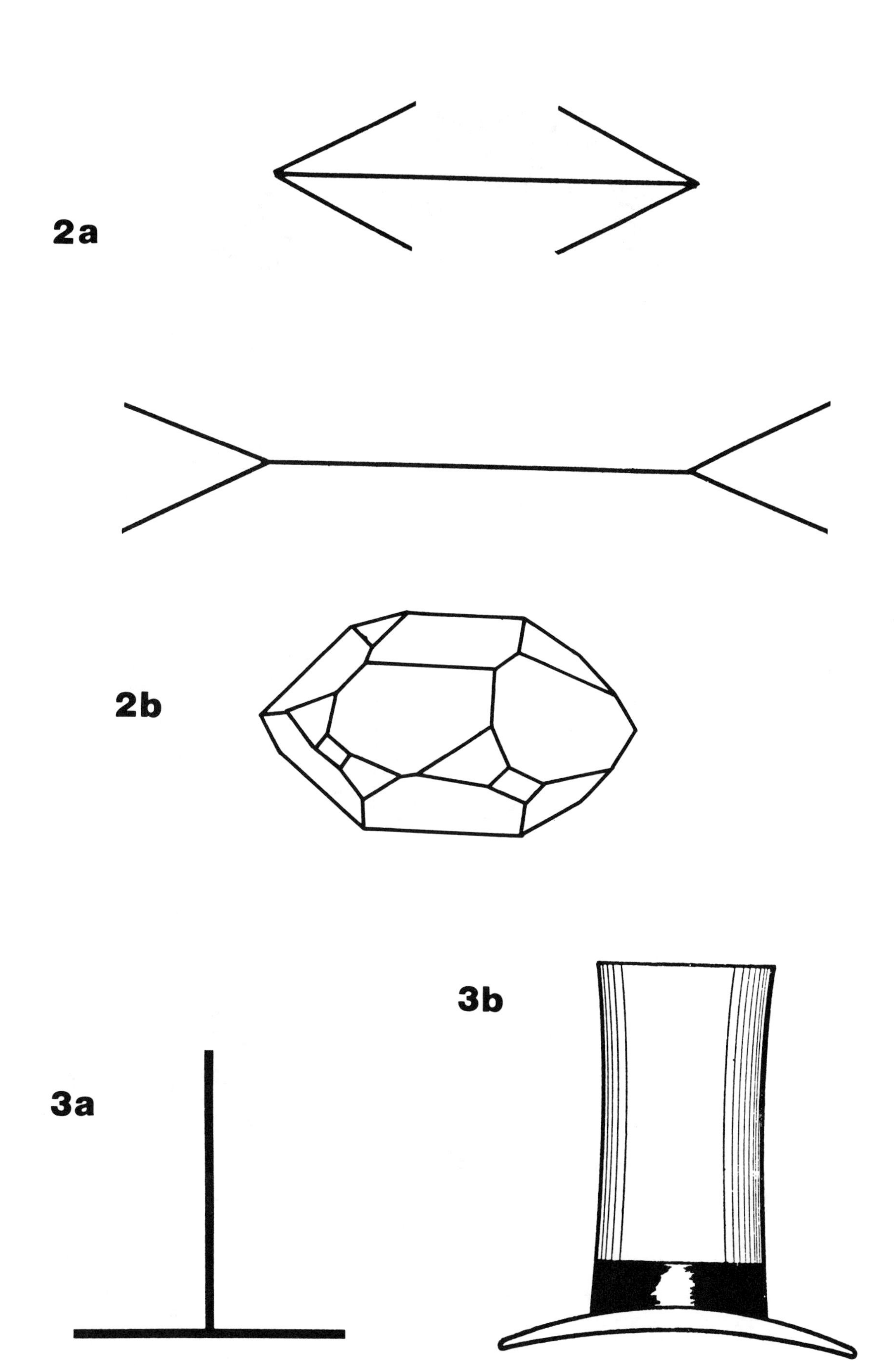
2a
2b
3b
3a

4

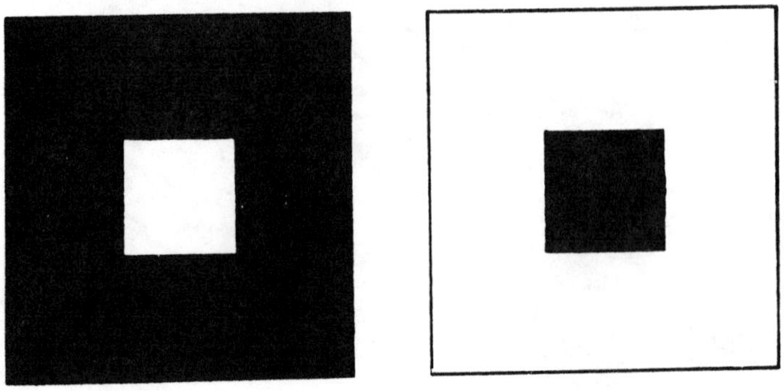

5a

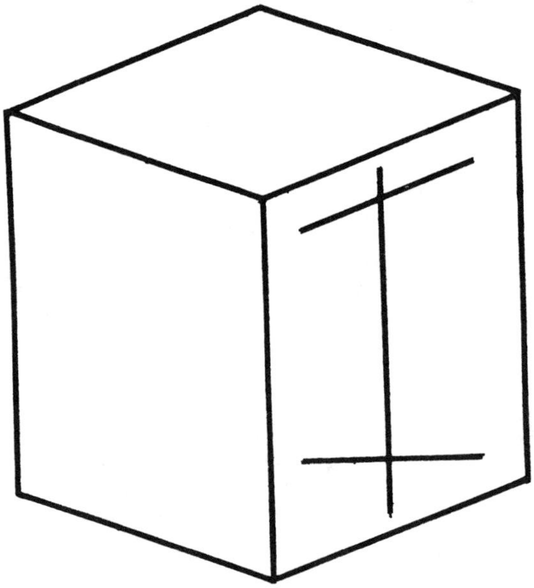

5b

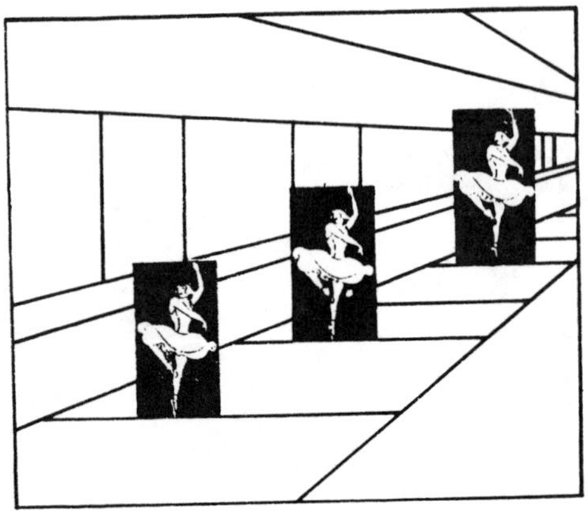

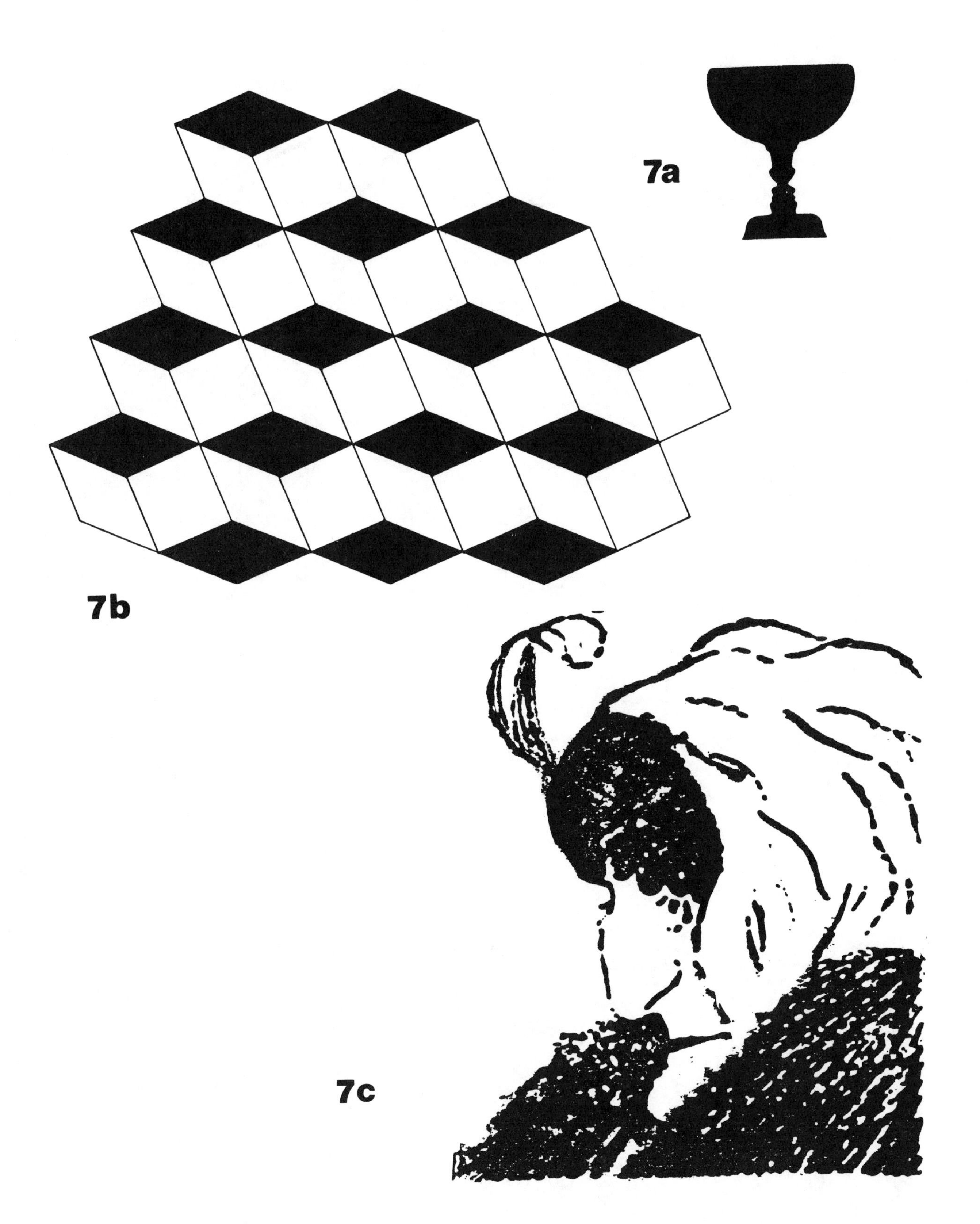
7a
7b
7c

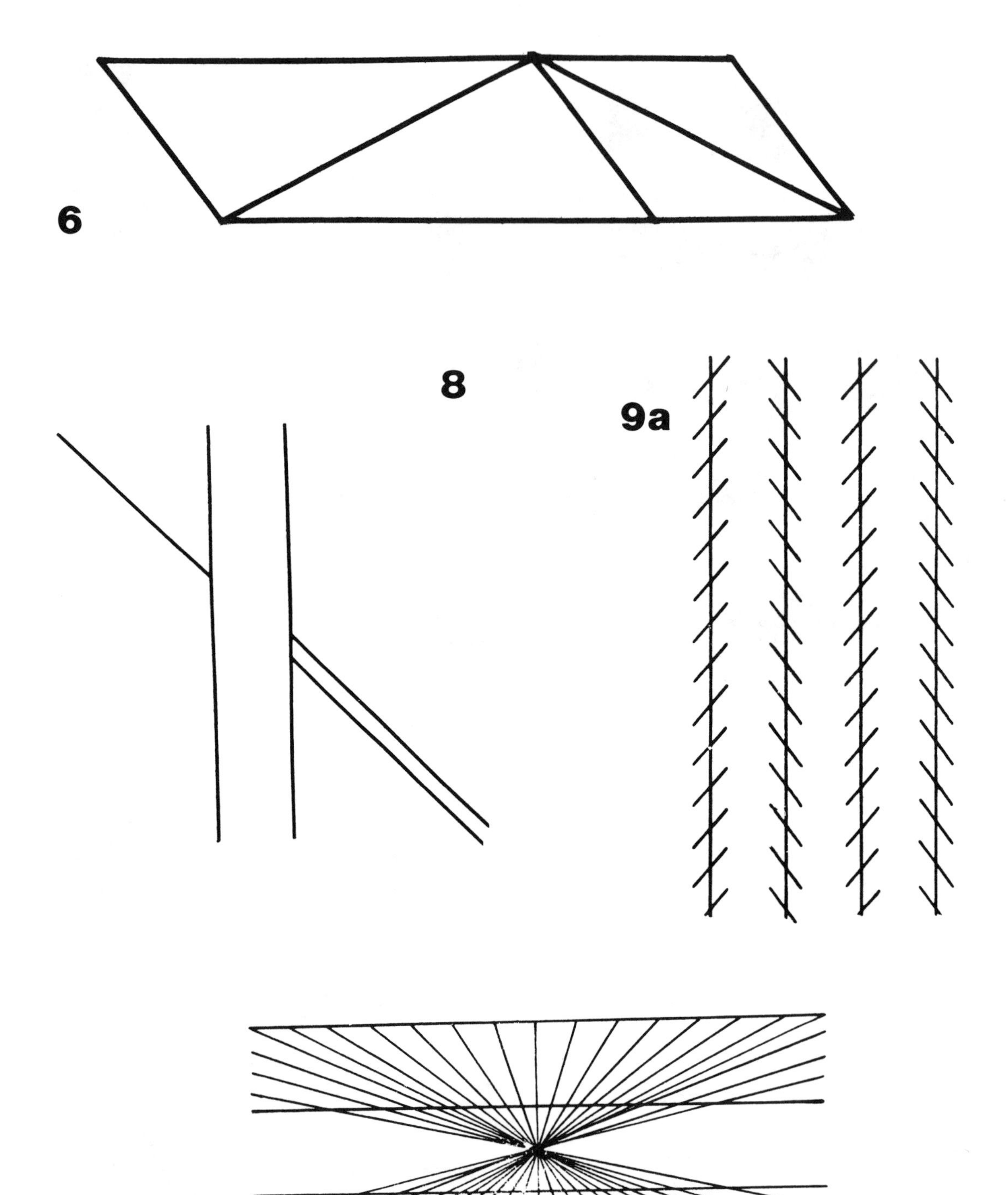

9b

10 a

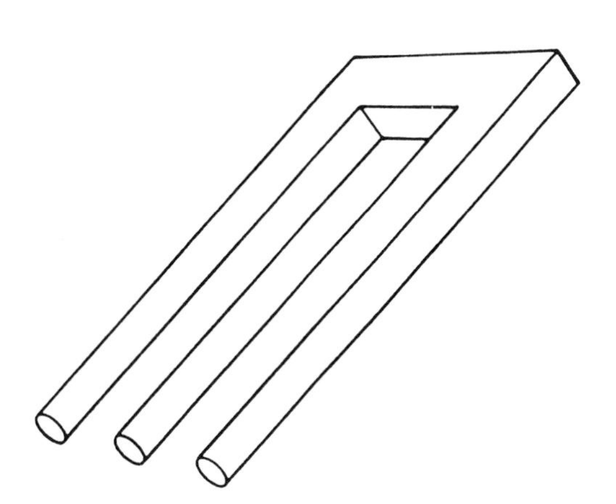

10 b

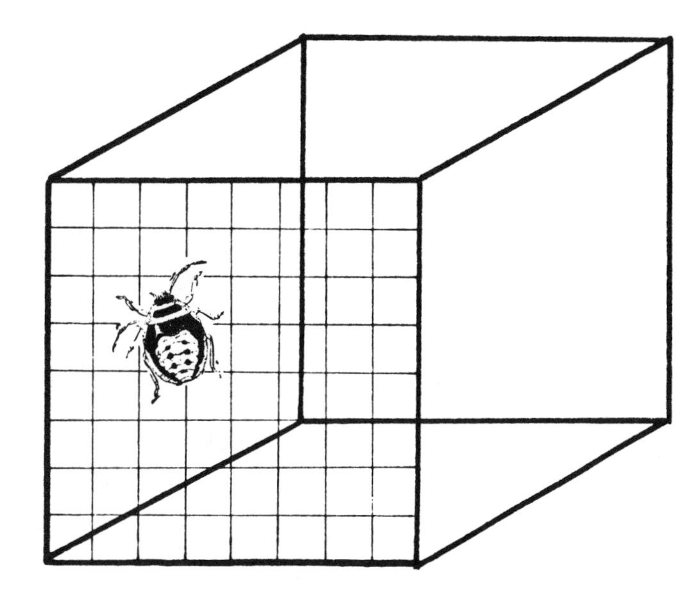

FLEXAGONS

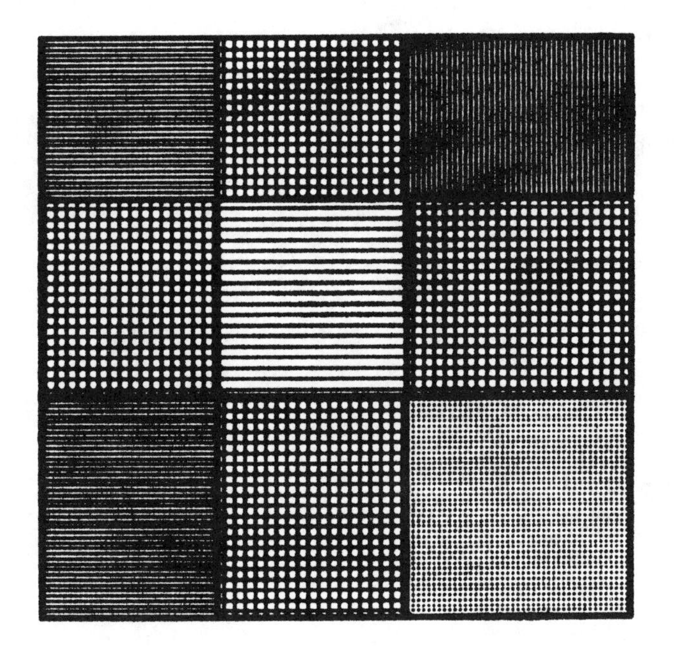

FLEXAGONS

BACKGROUND MATERIAL

In a broad sense flexagons can be considered a type of topological model. They are figures made from a sheet of paper, but end up having a varying number of faces which are brought to view by a series of flexing.

The flexagons probably first appeared as toys and magic devices. In fact, the tri-tetra flexagon, called *Jacob's Ladder*, was a popular toy in the 1890's. A picture of it appeared in *Magic: Stage Illusions and Scientific Diversions* by Albert H. Hopkins in 1897. Even today it is a popular toy going by the names *Klik-klak Box* and *Flip Flop Blocks.*

The creation of the hexa-hexa flexagon in 1934 by Arthur H. Stone actually initiated mathematical interest in these intriguing figures. At the time, Stone was a 23 year old English graduate student at Princeton University. In order to make his American sheets of paper fit his English notebook, he had to trim a strip off each sheet. He began fiddling with these strips of paper, folding them in various ways; and thus came up with the hexa-hexa flexagon. As a result, he and three friends, Bryant Tuckerman, Richard Feyhman, and John Tukey, studied the properties of hexa-hexa flexagons and developed a complete mathematical theory for them. Since then a number of scientific papers have been written on flexagons.[1]

TETRA FLEXAGONS

Tetra flexagons are 4-sided flexagons. The simplest of the tetra-flexagons is the **tri-tetra flexagon.** The **tri** indicates how many faces, and the **tetra** indicates how many sides in the figure. It is the tri-tetra flexagon with ribbon hinges which is found in magic and novelty shops, and is dubbed the *magic bill fold* (used to conceal a dollar bill).

tetra-tetra flexagons

The tetra-tetra flexagon has 4 faces and is a 4 sided figure. It can be made in a variety of ways. The one shown is the one most often illustrated.

1 Martin Gardner's article in *Scientific American,* 1956, O.C. Oakley and R.T. Wisner's article in the March 1957 issue of *The American Mathematical Monthly*

Making a tri-tetra flexagon:

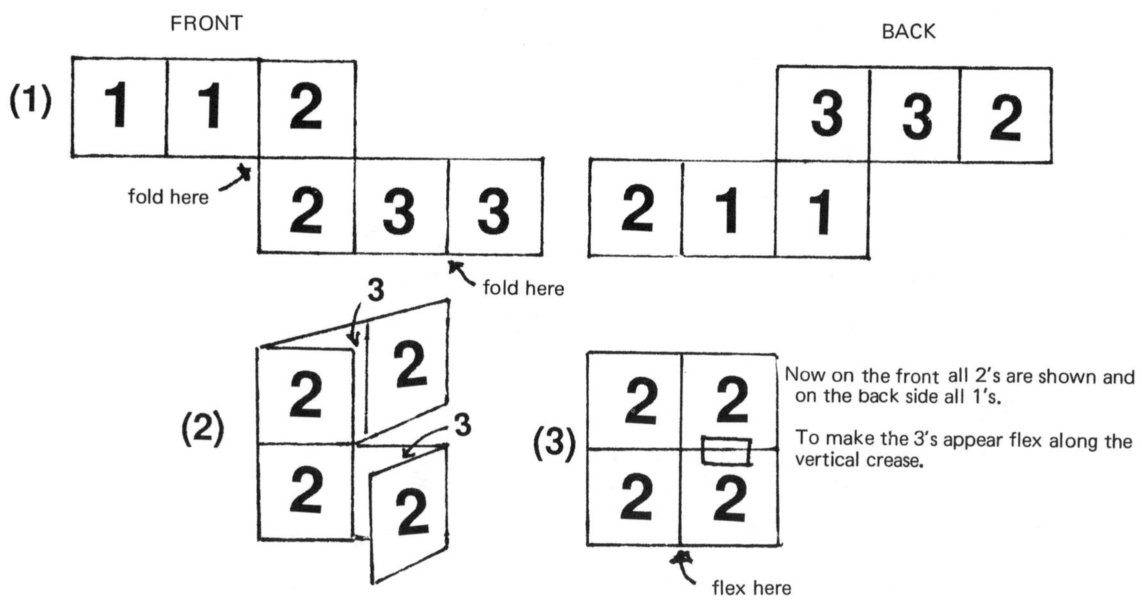

Making a tetra-tetra flexagon:

The tetra-tetra flexagon can also be made the same way but with 2 rows of squares. Its flexing properties are still preserved when one or more rows are added.

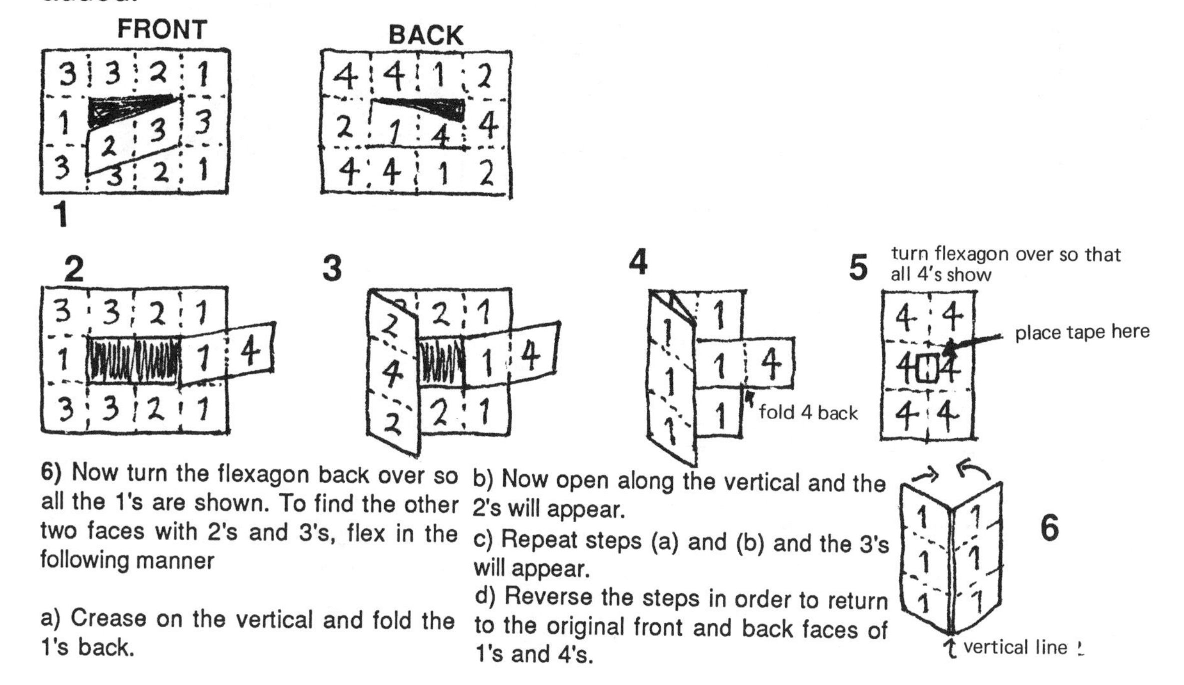

6) Now turn the flexagon back over so all the 1's are shown. To find the other two faces with 2's and 3's, flex in the following manner

a) Crease on the vertical and fold the 1's back.

b) Now open along the vertical and the 2's will appear.

c) Repeat steps (a) and (b) and the 3's will appear.

d) Reverse the steps in order to return to the original front and back faces of 1's and 4's.

Other higher order tetra flexagons can be constructed, for example the octo-tetra flexagon (8 faces and 4 sides).

Tetra flexagons with an even number of faces, e.g. hexa-tetra flexagons, are made from similar starting rectangular patterns. Tetra flexagons with an odd number of faces use patterns similar to the tri-tetra flexagon.

hexa-tetra flexagon

Another type of tetra flexagon is the hexa-tetra flexagon (6 faces and 4 sides). What is especially unique about the hexa-tetra flexagon is that it may be flexed either along the vertical or the horizontal axis. It is constructed in the following manner.

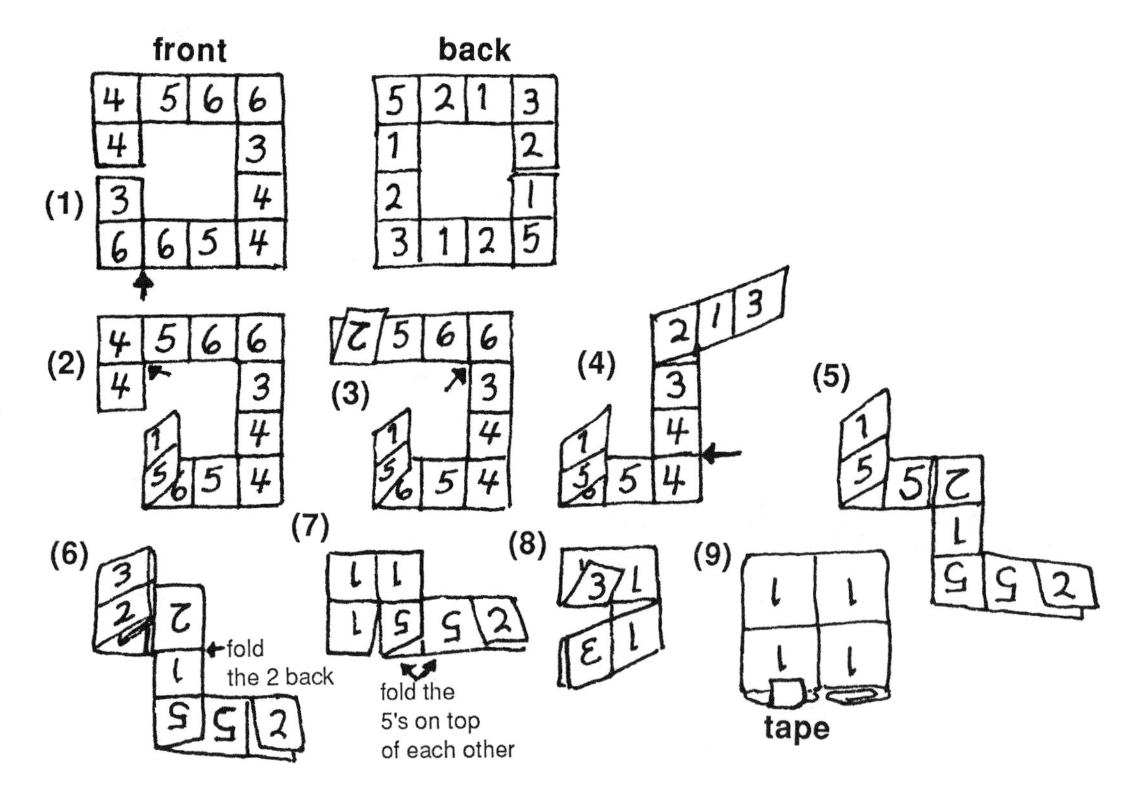

If large sized square strips are used, then higher order (more faces) tetra flexagons can be made. With this pattern, the order increases in increments of four, so the next tetra flexagon would be deka-tetra flexagon (four more faces than the hexa-tetra flexagon). Following this would come the 14-tetra flexagon, then the 18-tetra flexagons, and so forth. If one were to make a tetra flexagon of orders other than these, for example, dodeca-tetra flexagon (12 faces), other shaped strips must be used.

HEXA FLEXAGONS

The **hexa-hexa flexagon** was the one discovered by Stone in 1930. It has six faces and is six sided. It is formed from a single strip of paper and is constructed in the following manner:

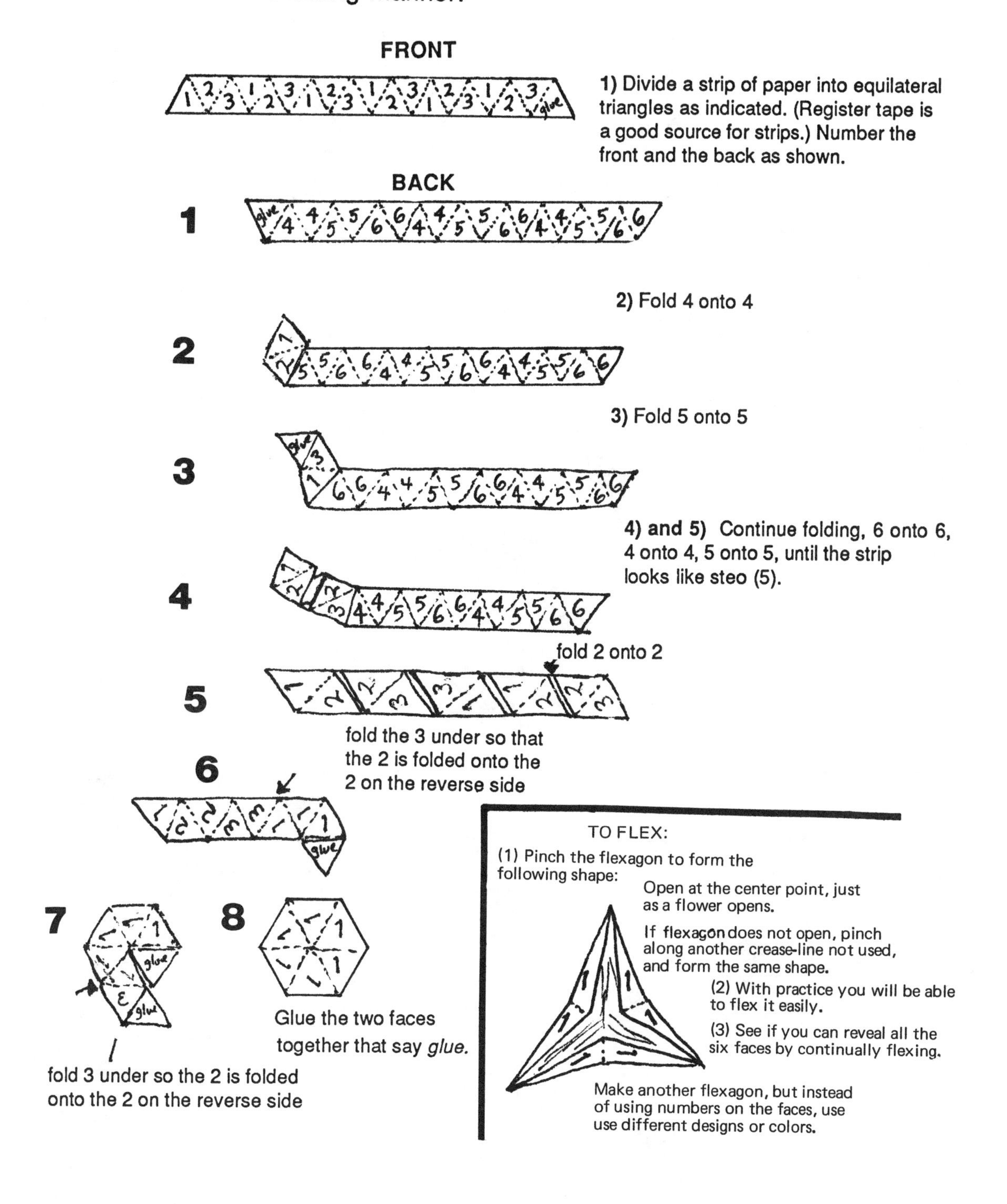

In the course of studying flexagons, Stone invented some fascinating objects, one of which was the tetra flexatube. It appears as a flat shaped flexagon, which opens into a tube. More amazing was that he discovered the tube could be turned inside out by flexing along the boundaries of a right angles.

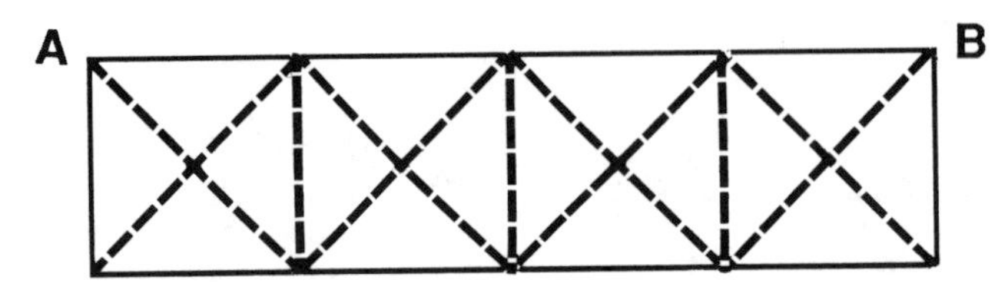

Pre-crease a rectangular strip of paper along the dotted lines, as indicated above.

Tape ends A and B together to form a flexatube – as illustrated below.

The problem is to turn the flexatube inside out by only folding along the crease marks.

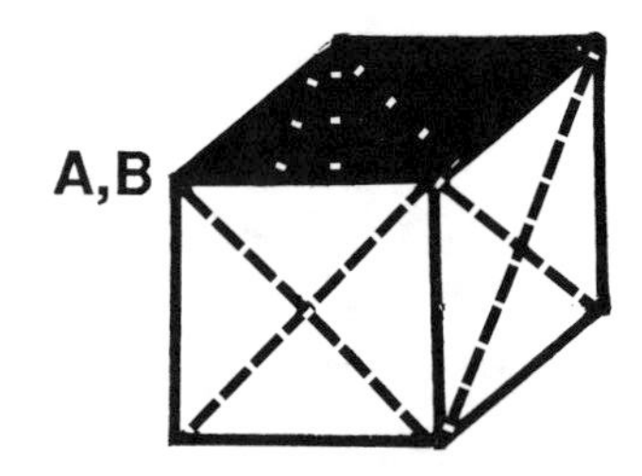

Flexagons originated as recreational novelties. Yet their fascinating properties prompted mathematicians to devote time and energy to discover new flexagons, new characteristics, and new properties. Although all their applications and uses have yet to be discovered or invented, flexagons still provide an interesting and intriguing manipulative and mental exercise.

LESSON

Show the class a long strip of paper that you will use to make a hexa-hexa flexagon, and show them a large hexa-hexa flexagon that you have previously made.

If time permits, decorate its six faces with either numbers, colors, and/or designs.

Illustrate its different surfaces and how they appear by flexing (it may be difficult to find all six faces without practicing ahead of time). You may want to hand it to a student to flex and discover various faces.

Referring to background material, relay Stone's discovery of the hexa-hexa flexagon.

At this point explain what a flexagon is, and present other completed flexagons illustrated in the background material.

Discuss how flexagons are named, as outlined in the background material. Emphasize that flexagons are mathematical models that have been studied and analyzed extensively, and whose applications have not been fully developed or discovered. They have, however, for over a hundred years brought enjoyment and intrigued. Point out that flexagons have been used as toys, magic tricks, and marketing gimmicks. Their mathematical applications have not been discovered entirely, although they have been studied by mathematicians for over forty years.

You will have to decide which flexagons you will have time to construct in class, and which flexagons you will have them construct on their own.

Bear in mind you will need to be sure the students are able to perform the flexing process, and are able to find the various faces of a particular flexagon.

LESSON'S OUTLINE

I. Introduction to flexagons
- A. Present strip and hexa-hexa flexagon
- B. Illustrate flexing on the hexa-hexa flexagon
- C. Historical background
  - 1) commercial use
  - 2) discovery of hexa-hexa flexagon
  - 3) flexagons as mathematical models

II. Constructing and flexing flexagons
- A. selection of flexagon to be constructed and demonstrated left to teacher's discretion.

III. Pass out assignment.

LESSON'S OBJECTIVES

1) Student will learn about a recreational form of mathematics that involves physical manipulation–called flexagons.

2) Student will be informed on the historical background of flexagons, their definitions and method of naming.

3) Student will construct and flex different types of flexagons.

4) Student will use the flexagons as a creative art form by developing designs on the flexagons constructed.

ANSWERS to assignment on flexagons

1) Student's model will serve as his or her solution.

2) a) 4, 10 b) 4, 12

3) Student's flexagon will be his or her solution.

4) Student's flexagon will be his or her solution.

5) Answers may vary.

ASSIGNMENT

1) Construct a tetra-tetra flexagon, following the diagram and instructions below.

FRONT

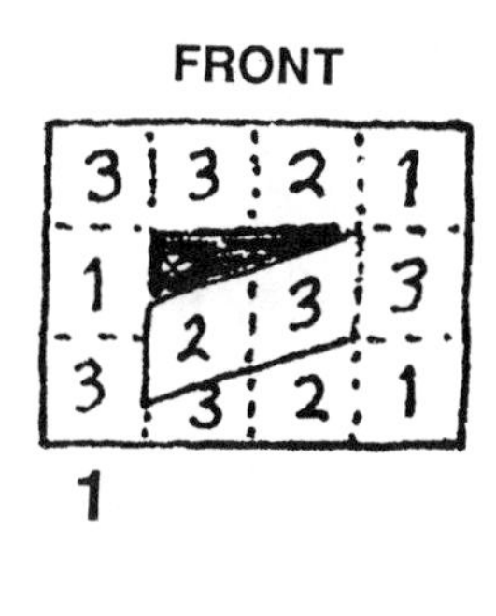

BACK

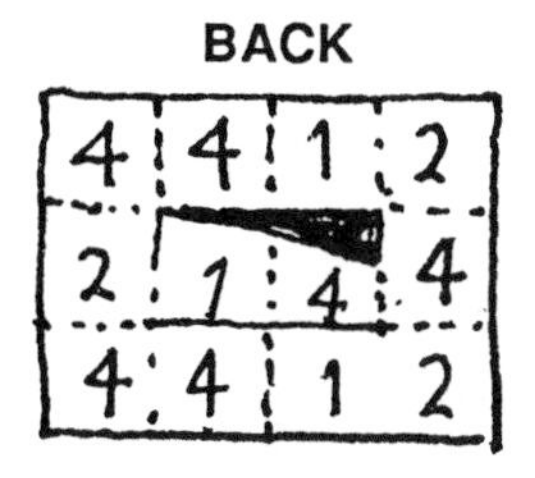

1

2

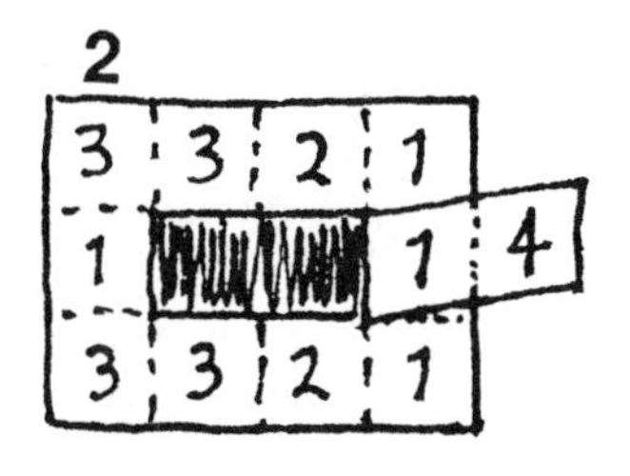

3

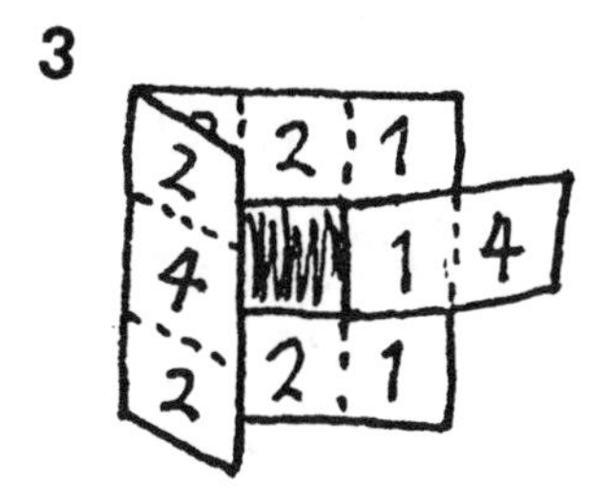

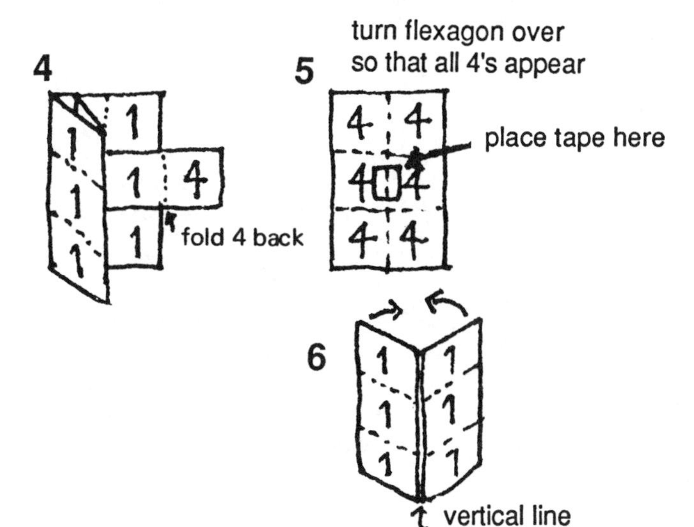

6) Now turn the flexagon back over so all the 1's are shown. To find the other two faces with 2's and 3's, flex in the following manner

a) Crease on the vertical and fold the 1's back.

b) Now open along the vertical and the 2's will appear.

c) Repeat steps (a) and (b) and the 3's will appear.

d) Reverse the steps in order to return to the original front and back faces of 1's and 4's.

2) How many faces and how many sides do the following flexagons have:

a) deka-tetra flexagon sides________ faces___________

b) dodeca-tetra flexagon sides_______ faces___________

3) Construct a hexa-hexa flexagon following the diagram below. Before folding and gluing, create your designs or patterns on both sides. Now fold and glue. Flex to see the designs you have created.

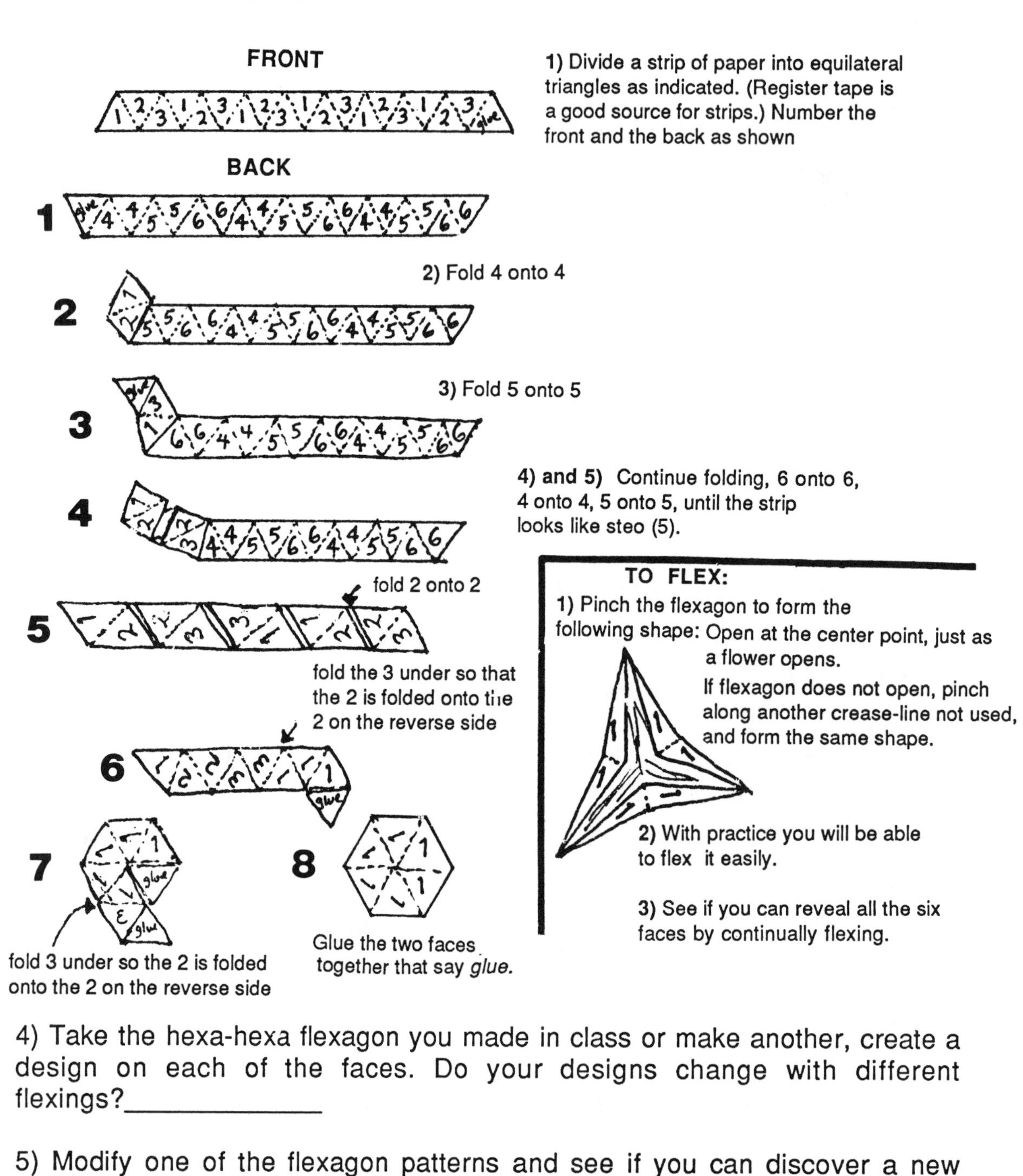

4) Take the hexa-hexa flexagon you made in class or make another, create a design on each of the faces. Do your designs change with different flexings?_____________

5) Modify one of the flexagon patterns and see if you can discover a new flexagon. If so, how many faces and how many sides does it have?________________________ How should it be named?

THE CYCLOID and PARADOXES

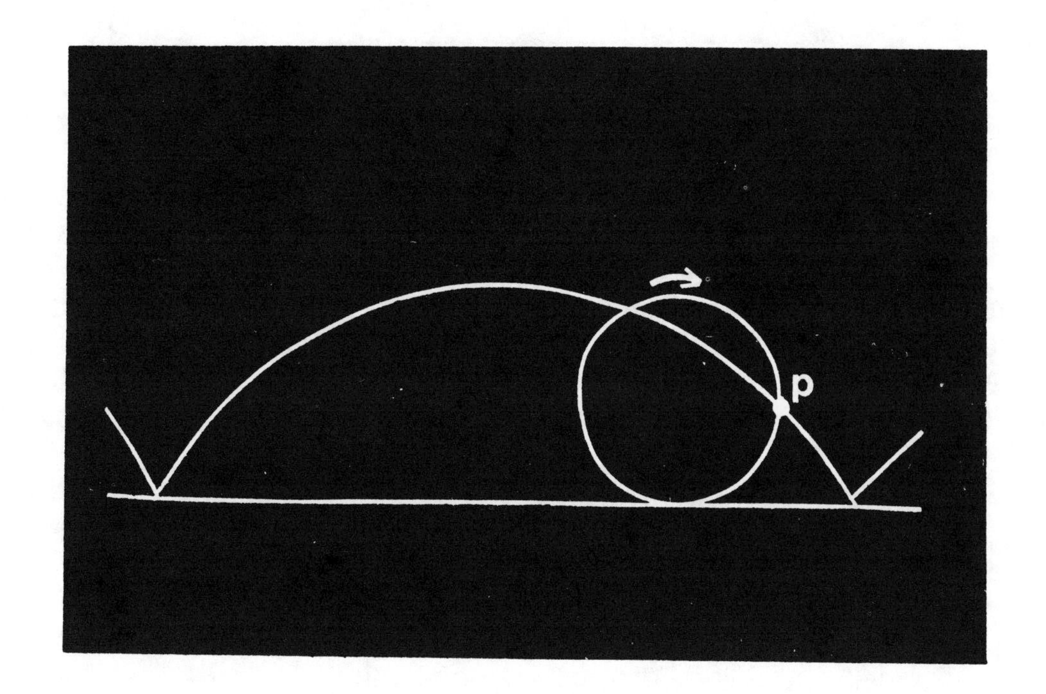

the cycloid and paradoxes

BACKGROUND MATERIAL

The *cycloid* is one of the many fascinating, yet controverisal curves of mathematics. It is *the curve traced by the path of a fixed point on a circle which rolls smoothly on a straight line.*

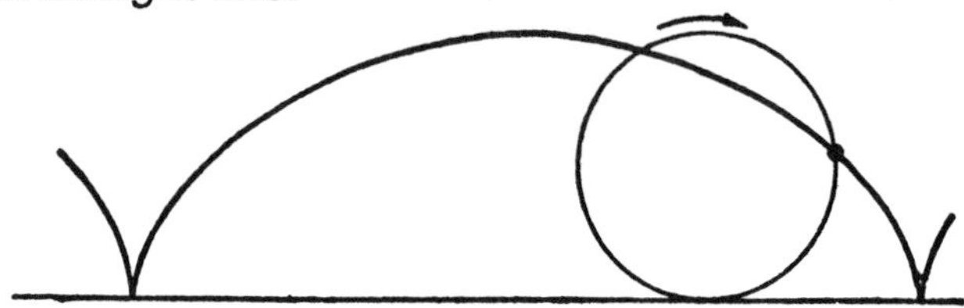

One of the first references to the cycloid appears in a book by Charles Bouvelles published in 1501. But it was in the 17th century that a large number of prominent mathematicians[1]were intent on discovering its properties. The 17th century was a time of interest in mathematics of mechanics and motion, which may explain the keen interest in the cycloid. Along with the many discoveries at this period of time, there were many arguments about who discovered what first, accusations of plagarism, and minimizations of another's work. As a result the cycloid has been labeled *the apple of discord* and *the Helen of geometry.* Some of the properties of the cycloid that were discovered during the 17th century are:

1) **Its length is 4 times the diameter of the rotating circle.** (Discovered in 1658 by Sir Christopher Wren.)

2) **The area under the arc is three times the area of the rotating circle.** (Proven independently in 1644 by Evangelista Torricelli and Gilles Personne de Roberval)

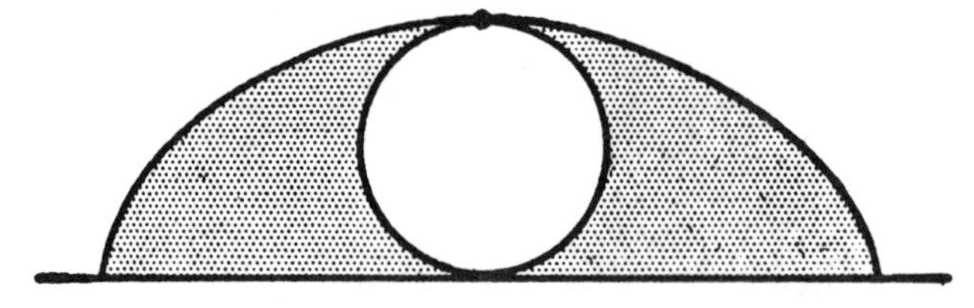

By property 2, this means that the shaded region on either side of the circle is equal to the area of the circle.

[1]These prominent mathematicians included: Galileo, Pascal, Roberval, Torricelli, Descartes, Fermat, Wren, Wallis, Huygens, J. Bernoulli, Leibniz, and Newton.

3) The point on the circle that is tracing the cycloid takes on different speeds–in fact, at one place it is even at rest. Point P on the top of the circle moving downward moves faster than when the point is on the bottom moving upward. (**note**: Reference here is to horizontal velocity of these points with relation to the ground.)

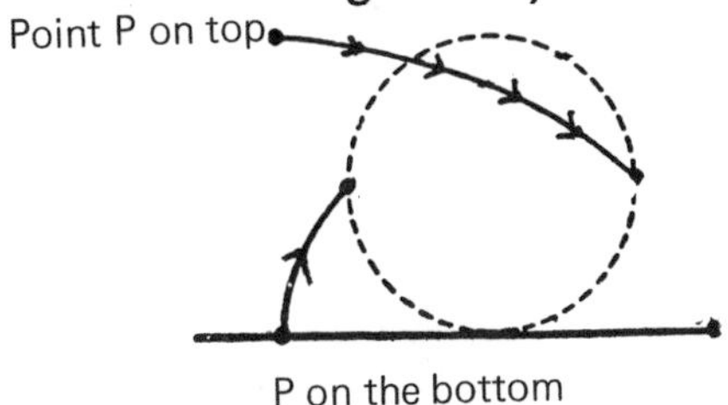

Point P on top moves faster in a quarter turn than point P on the bottom. This is because point P on top has gone further in a quarter turn than point P on the bottom (diagram1).

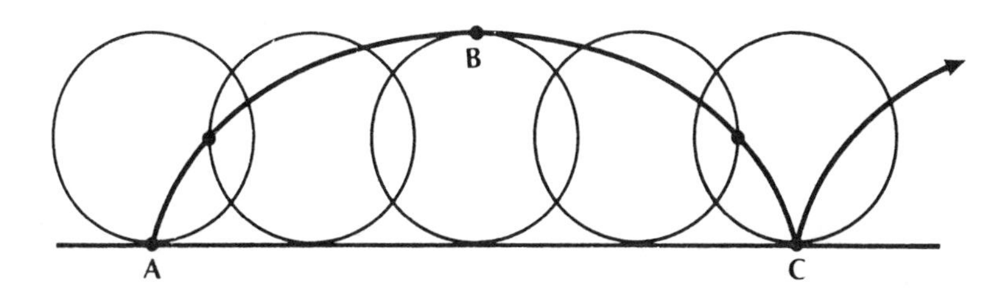

Point P is at rest when it lands on either A, B or C because it is changing its movement from downward to upward or vice versa.

4) When marbles are released from different points of a cycloid shaped container, they arrive at the bottom simultaneously.

Some interesting paradoxes which result from the properties of the cycloid are:

1) the coin paradox

The top coin has been moved halfway around the coin below it. Yet it ends up in the same position it originally started. Since it traveled half its circumference, one would expect it to end upside down.

2) rotating wheel paradox

The top of a rolling wheel moves faster than the bottom. This phenomenum is a consequence of the third property of a cycloid.

3) The larger circle makes one complete revolution, so that |AD| is the length of this circle's circumference. The smaller circle has also made one revolution and covered the distance |BC|. It is clear that |AD|=|BC|, but we know that the larger circle's

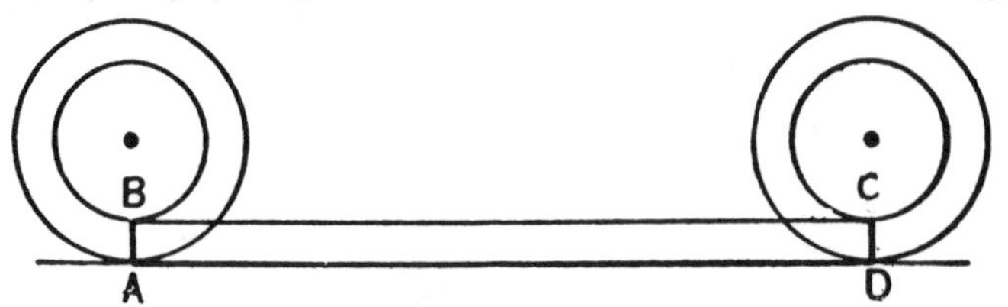

circumference is greater than the smaller circle's. To explain this paradox we need to look at the *prolate cycloid* – the path traced by an inner point of a wheel while it rolls in a straight line.

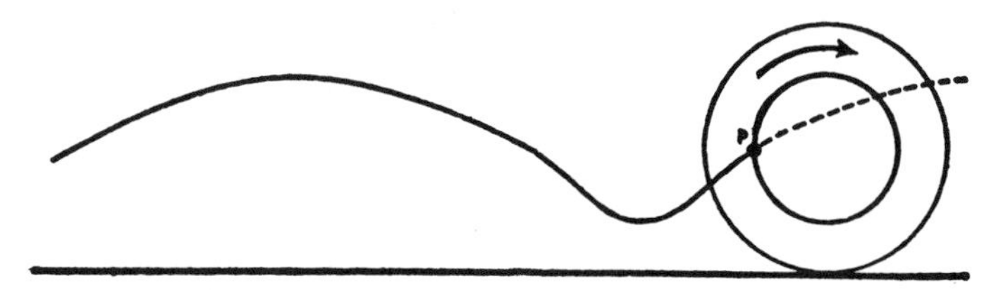

The small circle covers the distance |BC|, while it is rolling and part while it is being carried by large circle as it moves from A to D. Similarly this occurs to the center of a circle.

Another explanation is to consider the center of the circle as the smaller circle. The center of any circle is a point. Since a point has zero dimension, it does not actually revolve. Thus, the center is carried the entire distance from A to D by the large circle, although it has zero circumference.

Galileo's explanation: Galileo analyzed this problem by considering two concentric squares on a square "wheel". As the box flips 4 times, transversing the perimeter of the square "wheel", |AB|, we notice the small box is carried along 3 jumps. This illustrates how the small circle is also carried along the distance |AB|, and that |AB| does not represent its circumference.

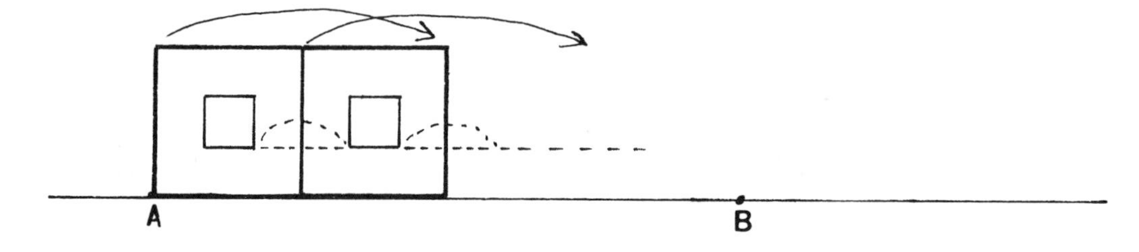

4) the train paradox

At any instant a moving train never moves entirely in the direction the engine is pulling. There is always part of the train moving in the opposite direction than that in which the train is moving.

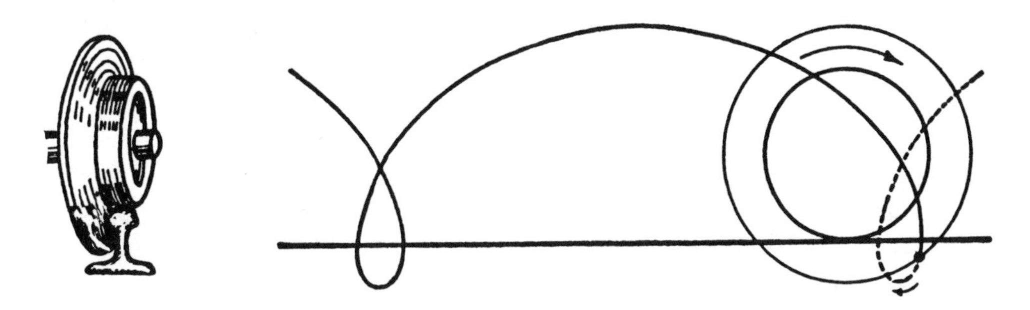

This paradox can be explained by using the cycloid. In this case, it is called a *curtate cycloid* – the curve traced by a fixed point outside a revolving wheel. The diagram shows the part of the train wheel that moves backwards as the train moves forward.

LESSON

An interesting way to introduce the cycloid is to first talk about paradoxes in general. Ask your class for a definition or an example of a paradox. One definition of a *paradox* is *a proposition which is logically valid but seems impossible or absurd, when it comes to common sense.*

Paradoxes are interesting, entertaining, and a very important part of mathematics. They emphasize how important it is to state and prove ideas carefully so there are no loopholes. In mathematics, we try to make mathematical ideas cover as many facets as possible, i.e. we try to generalize a concept and thereby try to make it apply to more objects. It is important to generalize, but it can be dangerous. One must proceed cautiously. Some paradoxes illustrate this danger.

Here are three paradoxes to introduce your class to the realm of paradoxes:

1) "ALL TEACHERS ARE LIARS"

This paradox is a take off of the famous *Liars Paradox* by the legendary poet, Epimenides.[2]

See if your students can arrive at the paradox, i.e. *since you are a teacher stating this, you are lying and if you are lying the opposite must be true–all teachers are truth tellers. But then that would make you a truth teller.* Thus we end up going around in a circle.

2) ZENO'S RUNNER PARADOX

Before a runner reaches the finish line the runner must pass the halfway point, then reach the 3/4 point (which is the halfway point of the remaining distance). For the last 1/4 to be completed another halfway point must be reached. These halfway marks continue indefinitely. The runner must pass an infinite number of halfway marks, and thus he will never reach the finish line.

The paradox can be explained by considering that between any two points there lie infinitely many points. The distance between two points is independent of the number of points between them, but depends on the length of the segment joining the two points.

[2] Epimenides was a Cretan. His paradox stated: "all Cretans are liars."

3) A VARIATION OF THE UNEXPECTED HANGING PARADOX[3]

A teacher announces that the test will be given on one of the five week days of next week, but tells the class, "You will not know which day it is until you are informed of your 1 p.m. test at 8 a.m. that day."

Why isn't the test going to be given?

The test cannot be on Friday because it is the last day it could be given and you could deduce this on Thursday, if you had not had it yet . And the condition was that you will not know which day until the morning of the test. So if it were not on Friday, that would make Thursday the last possible day. But it cannot be on Thursday because by Wednesday you would know there are Thursday and Friday left. Since Friday was out, on Wednesday you would know ahead of time it would be Thursday, which you are not suppose to know ahead of time. Now this leaves Wednesday as the last possible day, but Wednesday is out because if you did not have it by Tuesday, you would know by Tuesday you would have it on Wednesday. Continuing with this reasoning, each day of the week is eliminated.

With these examples laying the foundation as to the confusion paradoxes can cause, the **rotating wheel paradox** should be introduced (i.e. the top of a moving wheel moves faster than the bottom)–see background material for details. A physical way to present this paradox is to tape a circular piece of paper to the end of a large can with spokes drawn on the circular paper. Now roll the can and see that the students keep their eyes fixed–not moving with the can. Notice that the top spokes appear blurred while the bottom spokes appear distinct.

To explain this paradox and the other wheel paradoxes from the background material, tell your class that "We'll need to look at an unusual curve called a ***cycloid*** ." Give the defintion of the cycloid and the history of it by referring to the background material. Use the included diagrams to makes transparencies for the overhead projector.

One diagram shows that the distance covered from the top location of point P through a quarter turn of the circle is greater than from the bottom location of point P through a quarter turn. Since the time it takes the wheel to make a quarter turn is the same, that means the one covering the most distance is the faster, i.e. from the top point P. This explains why the spokes on the top appear blurred.

[3] *The unexpected hanging paradox* appeared in an article by Michael Scriven in the July, 1931 issue of the British journal, ***Mind***.

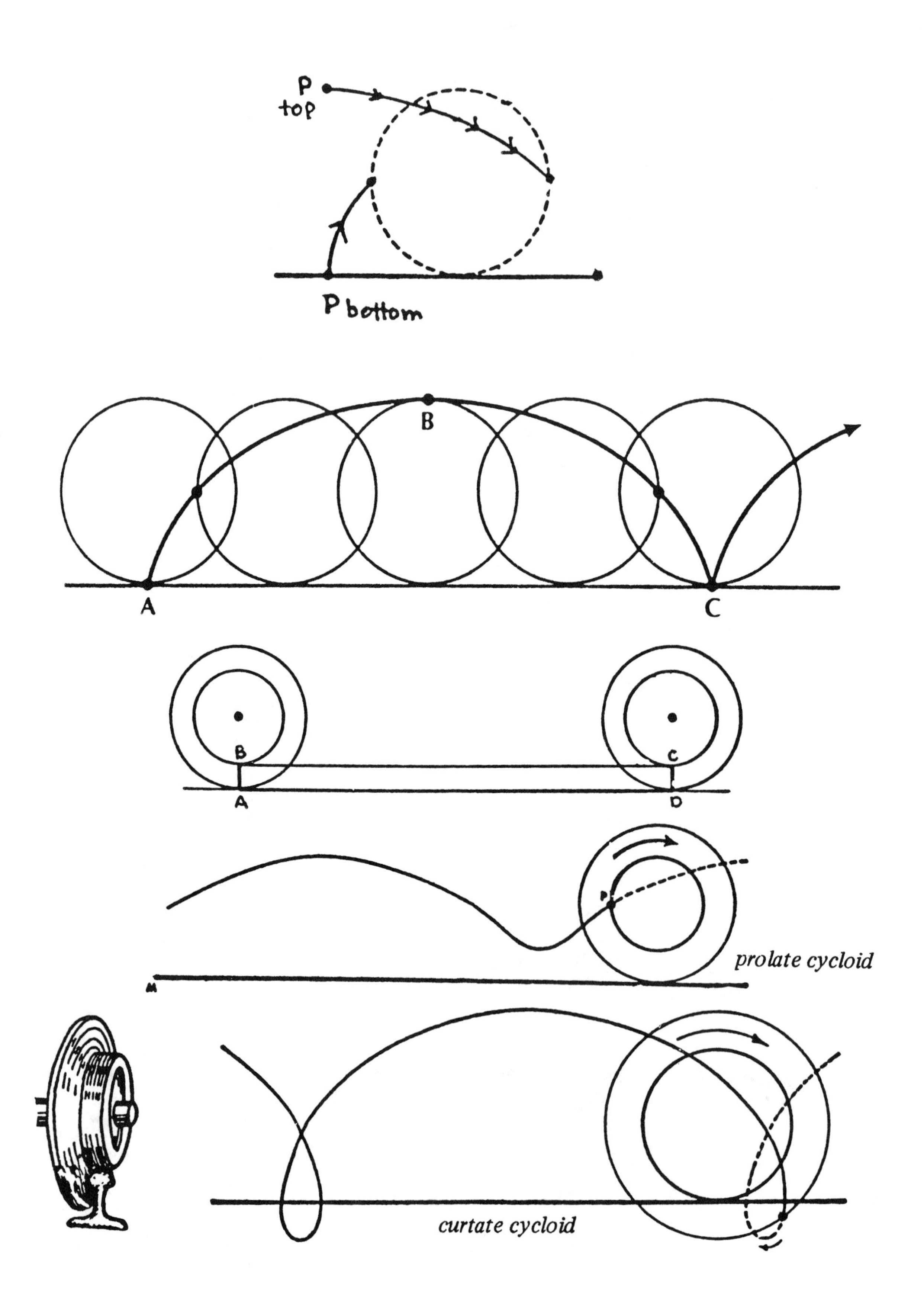
P
top
P bottom
B
A
C
B
C
A
D
prolate cycloid
curtate cycloid

With this introduction to the cycloid, present the **coin paradox**. Illustrate it on the board or with coins on an overhead projector. Let students have enough time to think, consider, and perhaps arrive at an explanation.

Referring to the background material now consider the ***prolate cycloid paradox*** with your class. Again use the diagram to make a transparency for the overhead projector, unless you prefer to draw the diagram on the board.

Finally, end with the **train paradox** using the ***curtate cycloid***, as covered in the background material.

Conclude by discussing some of the properties of a cycloid that were discovered during the 17th century (see background material). According to your class' math background, use your judgement as to the details and explanations of these properties.[4] Taylor them to your class' needs.

Emphasize that there is not one type of paradox and that you have discussed a variety of types. You will also introduce them to some other paradoxes in the enclosed assignment.

[4] Advanced classes can be asked to prove these properties.

LESSON OUTLINE

I. Develop the meaning of paradox

- A. Discuss why the study of paradoxes is important in mathematics
  - 1) the danger of generalizing
- B. Present 3 classic paradoxes
  - 1) "All teachers are liars" paradox
  - 2) Zeno's runner paradox
  - 3) the unexpected test paradox

II. Cycloid

- A. Introduction–the paradox, the top of moving wheel moves faster than the bottom of the wheel
- B. Define cycloid
  - 1) use cycloid to explain II.A.
  - 2) coin paradox
  - 3) train paradox

III. Summary of key points

IV. Pass out assignment.

LESSON'S OBJECTIVES

1) The student will be introduced to paradoxes and learn why they are important in the study of mathematics.

2) The student will learn what a cycloid is and the historical background on its discovery.

3) Student will learn about three paradoxes caused by the cycloid, the prolate cycloid, and the curtate cycloid and why they occur.

4) Student will be introduced to some of the properties of a cycloid, i.e. its length is 4 times the diameter of the rotating circle and the area under the arc is 3 times the area of the rotating circle.

5) Students will discuss three famous paradoxes–the liars paradox, Zeno's runner paradox, and the unexpected hanging paradox.

ANSWERS to the cycloid and paradoxes assignment

1) a) 40 inches b) 75π square inches

2) They arrive simultaneously by property (4) from the background material.

3) Square's area is 64 square units while the rectangles area is 65 square units. The pieces do not actually fit perfectly along the diagonal of the rectangle. There is a gap which totals 1 square inch in area.

4) This figure does not exist in our world, yet because we are viewing it our mind feels we must make sense out of it. Frustrated, our eyes search back and forth for meaning.

5) If the statement were false, then he would be hanged–but that would make the statement true. If the statement were true, then he would be shot–but that would make the statement false.

6) If you assume that "This sentence is false" is a true statement, then it would have to be false. But the statement is supposed to be true. Thus paradox leads the reasoning in a continual circle.

7) If the statement were true, then the visitor would be allowed to pass, but at the same time he would have to be hanged. Now if the statement were considered false, then the visitor would be hanged, but that would make the statement true.

8) If the barber shaves himself, he is breaking the rule that the barber shaves only those men in the village who do not shave themselves.

9) The symmetry that exists in the letters of this word follow the inversion property of a mirror.

10) "Point to the road that leads to the city which you are from." Thus both the liar and the truthteller will point to the city of truth.

ASSIGNMENT

Introductory information: The circle is one of the earliest curves familiar to humans. Its properties were known in ancient times. In fact, the appearance of pi, π,dates back 3000 years. π is defined as the circumference/diameter – the circumference of a circle divided by its diameter. π cannot be exactly represented by a decimal or a fraction, and therefore is an irrational number. Even more intriguing is the fact that the cycloid's arc length (formed by one revolution of a circle) can be represented by a rational number (that is, can be written as a fraction[5]).

properties of a cycloid:
1) The arc 's length is 4 times the diameter of the rotating circle.
2) The area under the arc is 3 times the area of the rotating circle.
3) The top of the circle moves faster than the bottom.
4) When marbles are released from different points of a cycloid shaped container, they arrive at the bottom simultaneously.

problems

1) If the circle used to generate a cycloid has radius 5 inches, then
   a) find the length of one arc of this cycloid

   b) find the area under the arc of this cycloid

2) If a metal is shaped into the form of an inverted cycloid solid, and two marbles are released simultaneously from point A and B, which marble will arrive at the bottom first and why?

3) The mathematician Charles Dodgson, whose pen name was Lewis Carroll (author of *Alice in Wonderland*) entertained friends with various mathematical puzzles. This geometric paradox was one of his favorites.

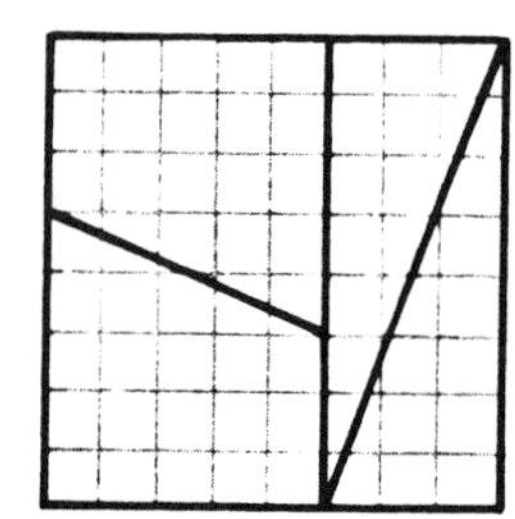

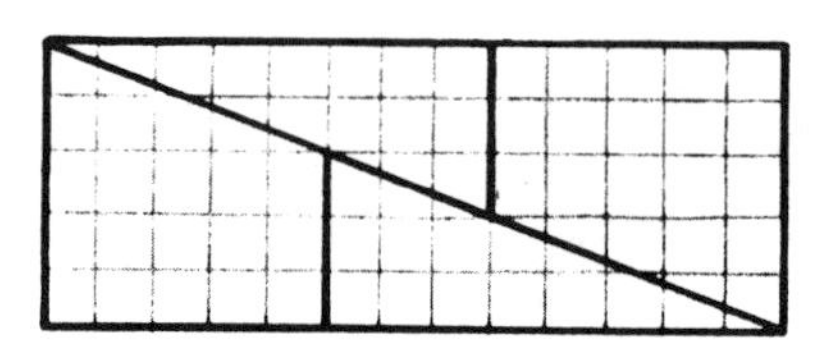

[5] A fraction or a rational number is defined as a number which can be put into the form, a/b, where a and b are integers with $b \neq 0$.

Find the area of the square pictured.

Now find the area of the rectangle that was formed from the pieces of this square.

*What is the result?*________________*Can you explain what happened?*

____

____

4) The optical illusion paradox. Explain what you see? ________________

____

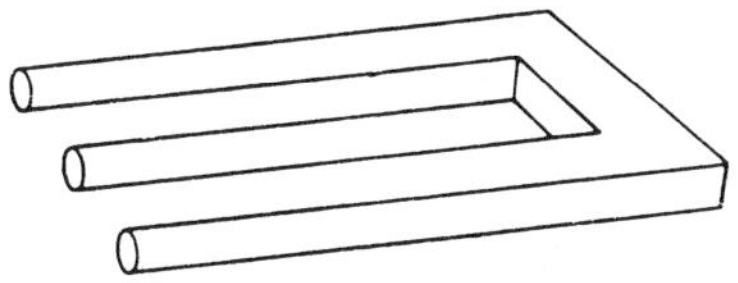

5) During the period of time known as the Wild West, people caught horse rustling were hanged. In this one particular town, the sheriff allowed the rustler to make a statement–if the statement were false, he would be hanged; if the statement were true, he would be shot to death. One logical rustler stated, "I shall be hanged." Why would this statement save him?

____

____

____

6) Consider the paradox created by the sentence, "This sentence is false." Explain how the sentence changes when you assume it is true.

____

____

____

7) The following paradox appears in *Don Quixote* chapter 51.

A guard asks all visitors to the city why they have come to visit the city. If they answer truthfully, he allows them to pass. If they lie, he hangs them. One day a certain visitor came and answered the guard's question with the statement, "I came to be hanged."

Explain the paradox by first considering the statement false, and tell what would happen. Then consider the statement true, and tell what would happen.

____

8) The *barber paradox* was first presented by Bertrand Russell, but it was attributed to an unnamed source in 1918.

the barber paradox: *In a certain village there is a barber. The barber shaves all and only those men in the village who do not shave themselves.*

question: Who shaves the barber? Explain the resulting paradox?

____

____

____

9) Hold the following word up to a mirror. What happens? __________

____

Explain why this happens. ____________________________

____

DIOXIDE

10) A traveler comes to a fork in the road. One path leads to the city of truth where everyone always tells the truth. The other path leads to the city of lies, where everyone always tells lies. A native from one of these cities is at the fork, but the traveler does not know from which of the cities he is. The traveler wants to know which path leads to the city of truth, but he does not know if the native will tell him the truth or a lie. He can ask only one question. What question could he ask? ____________________

____

TOPOLOGY

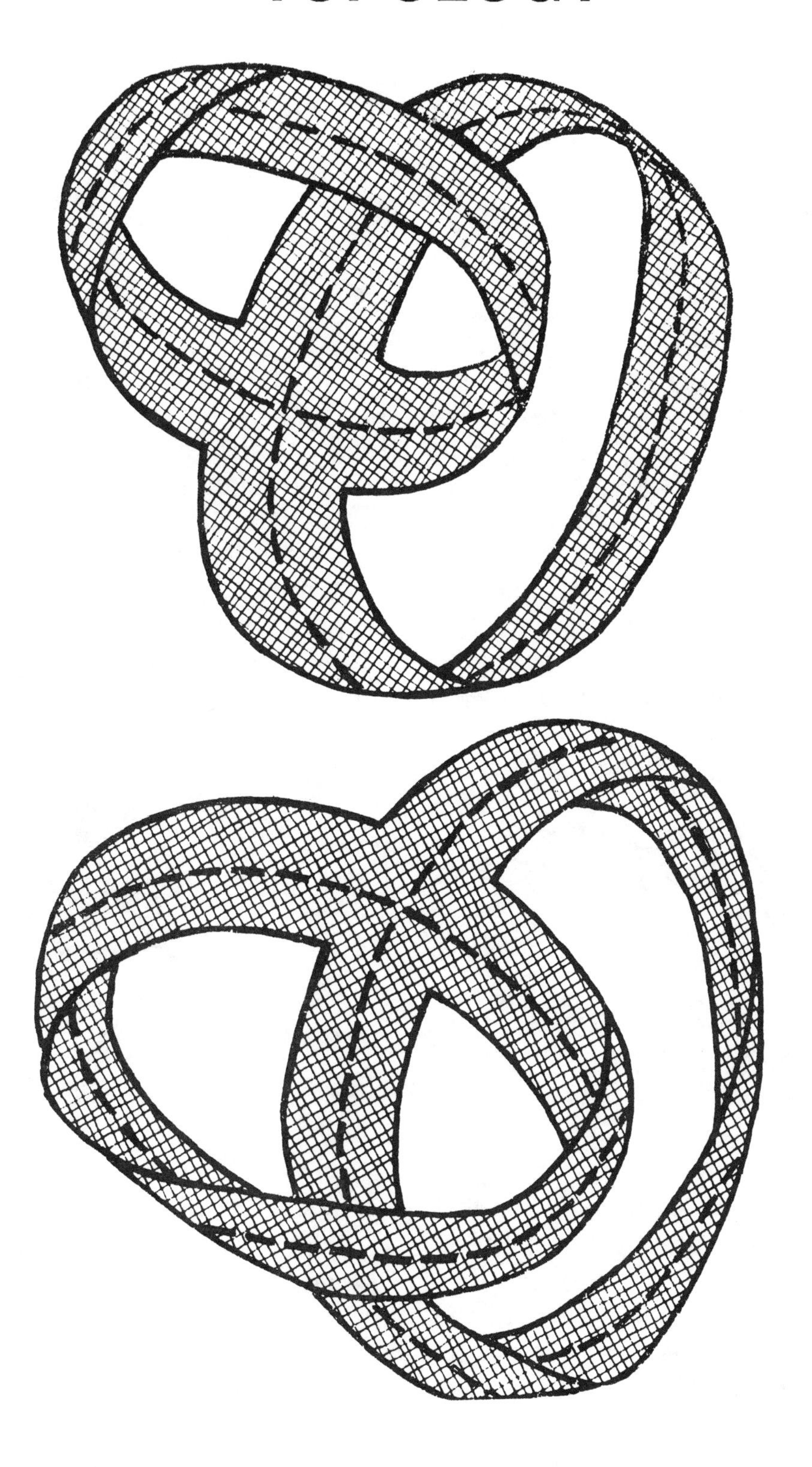

TOPOLOGY

BACKGROUND MATERIAL

Topology is a special kind of geometry. At first glance it seems like a conglomeration of varied, unrelated, and fascinating ideas with little practical applications–*the Moebius strip, networks, the Klein bottle* .

Topology originated with the solution in 1736 of a famous problem–the Koenigsberg Bridge Problem . Koenigsberg is a city on the Preger River that contains two islands and is joined by seven bridges. The river flows around the two islands of the town. The bridges run from the banks of the river to the two islands in the river with a bridge connecting the islands. It became a town tradition to take a Sunday walk, and try to cross each of the seven bridges only once. No one had solved the problem until it came to the attention of the Swiss mathematician Euler. At that time, Euler was serving the Russian empress Catherine the Great in St. Petersberg. In the process of solving this problem, Euler invented the branch of mathematics known as topology. He solved the *Koenigsberg Bridge Problem* by using an area of topology today called networks. A *network* is basically a diagram of a problem. The network for the Koenigsberg Bridge Problem is illustrated below.

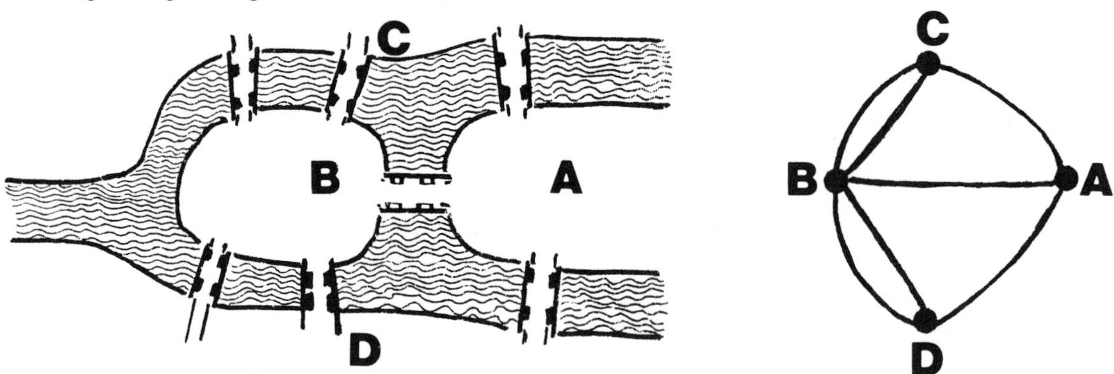

A network consists of *vertices* and *arcs* . A network is *traveled* or *traced* by passing through all the arcs exactly once. A vertex may be crossed any number of times. The above diagram shows the vertices for the Koenigsberg Bridge Problem as A, B, C, D. Note the number of arcs passing through each vertex–A has 3, B has 5, C has 3, and D has 3. Since these are all odd numbers, these vertices are called *odd vertices.* An *even vertex* would have an even number of arcs passing through it. Euler discovered many properties about the number of odd and even vertices a network can have and still be traceable. Specifically, Euler noted that for a vertex to be odd, one would have to begin or end the journey at that vertex. With this in mind, he reasoned that since a network can have only one beginning and only one ending, that a traceable network could have only two odd vertices. Thus, since in the Koenigsberg Bridge Problem there are four odd vertices, it cannot be traced.

This problem and Euler's solution launched the study of topology. Topology is a relatively new field. The mathematicians of the 19th century began delving into topology along with their studies of other non-Euclidean geometries. The first treatise of topology was written in 1847.

The definition of topology and some of its properties, concepts and objects are discussed on the following pages.

LESSON

Introduction: How many of you were ever asked to trace a shape similar to this, without lifting your pencil or doubling back? You were probably unaware

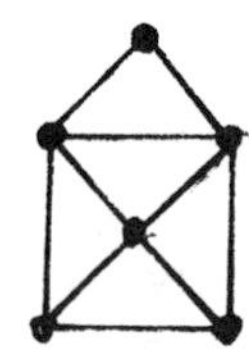 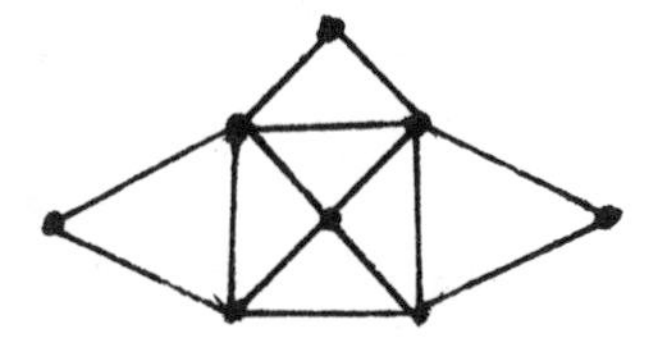

that you were solving a topological problem. In fact, this type of problem initiated the study of topology, in the 18th century (see background material on the Koenigsberg Bridge Problem).

Present the Koenigsberg Bridge Problem to the students and work through its solution with students. Include in your discussion the historical background of the problem.

To maintain the students' curiosity, introduce them to other types of topological problems shown below. I suggest you merely present the problems, and do not linger to solve them since they can be considered later. Mainly use this presentation of problems to stimulate interest.

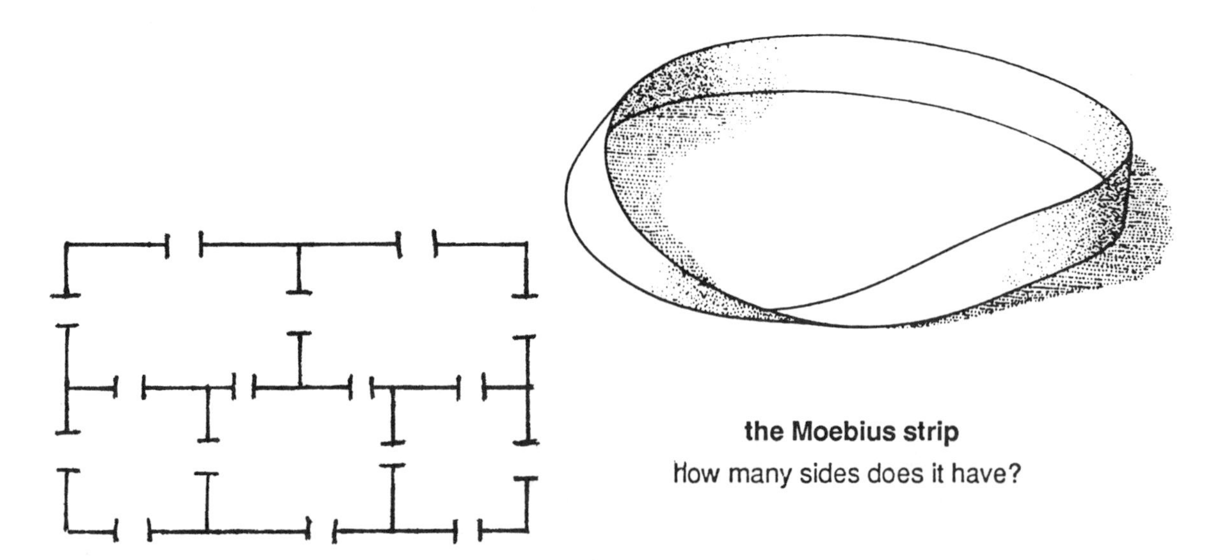

the Moebius strip
How many sides does it have?

This is the floor plan of an office.
Is it possible to pass through
each and every doorway only
once when making an inspection?

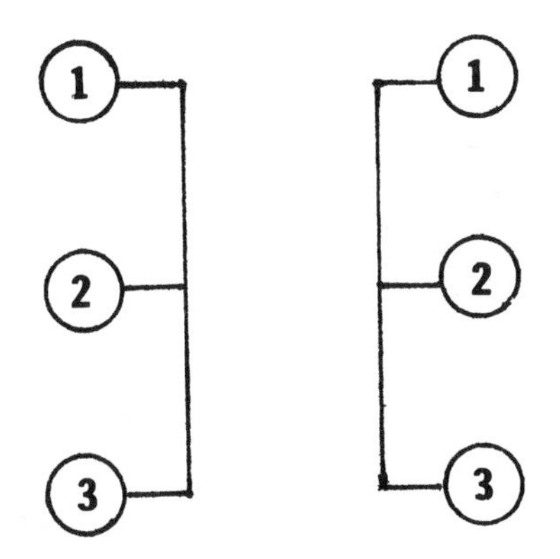

the caliph's problem
Match up like numbers without crossing any paths on the plane

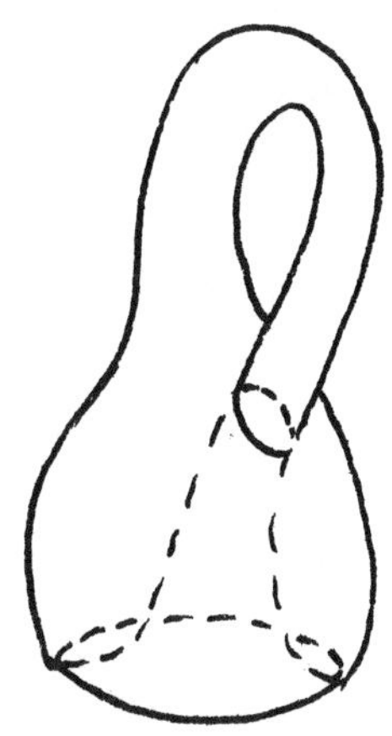

the Klein bottle
What happens when water is put in this bottle?

Now that students have seen a variety of topological problems, it would be an opportune time to discuss what the study of topology entails.

In topology, all these objects are considered the same or *equivalent* . The reason these are equivalent lies in the definition of topology.

Topology is a special kind of geometry that studies the properties of an object that remain unchanged when the object is distorted. By distorted we mean it can be stretched or shrunk.

Unlike Euclidean geometry, ***topology does not deal with size, shape, or rigid figures.***

Basically, topology studies *elastic figures* . This is why it is often referred to as ***rubber sheet geomerty*** **.** Imagine the objects of topology as existing on a rubber sheet that can be stretched and shrunk; and in the process of these transformations, we determine what characteristics remain unchanged. Thus in topology we never ask how long, how far, or how large ; but we ask *where, between what, inside* or *outside* .

examples: *Besure to give students time to consider these questions.*

A B C D

What properties of the figure remain unchanged here?

A B C D

Although the line is no longer straight, the points remain in the same position.

What properties remain unchanged?

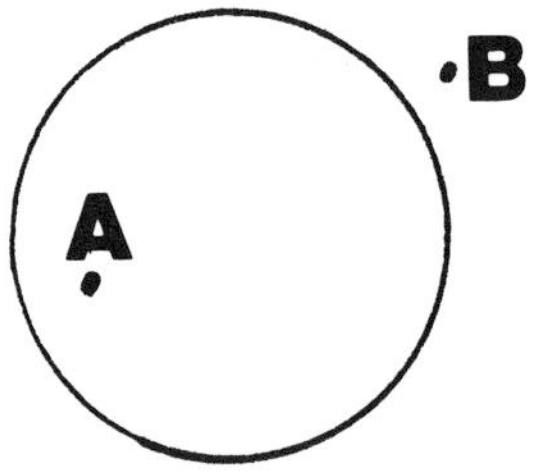

Here A remains inside and B is outside.

Thus, we see topology deals with *position* or *location* rather than shape and size.

Also in topology *distance has no meaning.*

The distance between A and B does not remain constant.

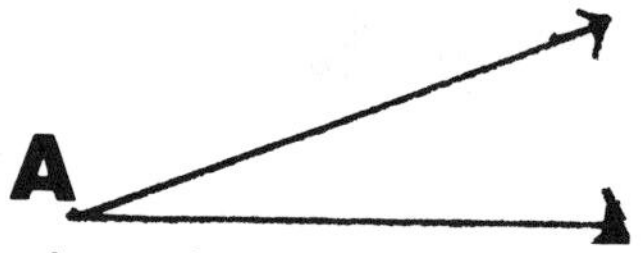

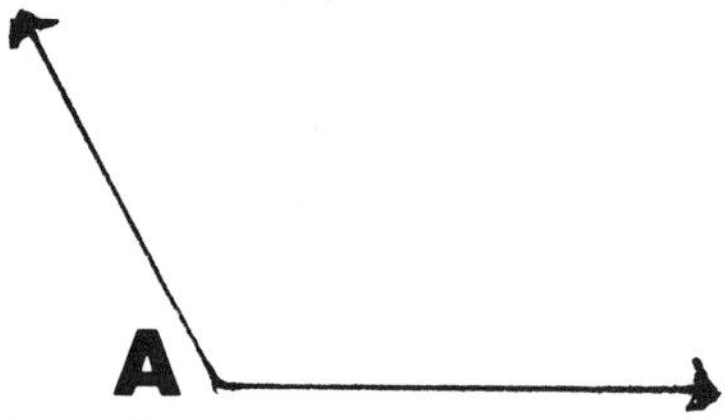

The size of an angle also has no meaning in topology.

Thus we see there are *no rigid objects* in topology, rather they can change *size, shape* and *position* .

How topology categorizes or identifies objects, since size and shape are unimportant.

Classification of objects was first done by French mathematician C. Jordan in the 19th century when he defined a *simple closed curve* as one which separates a plane into one inside region and one outside region.

As a result these objects are considered *equivalent simple closed curves.*

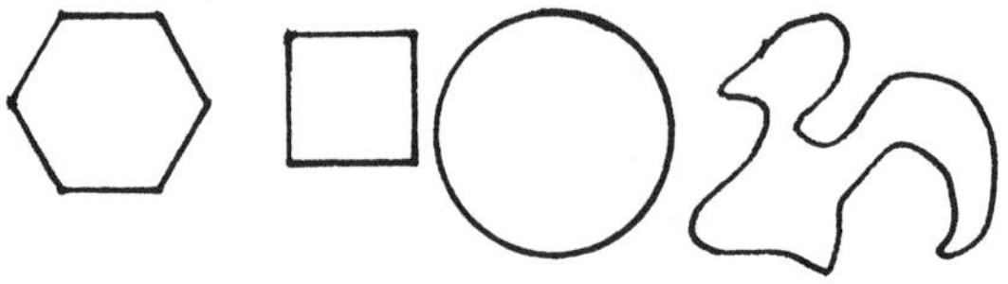

Ask the students why the curves below would not be considered simple closed curves.

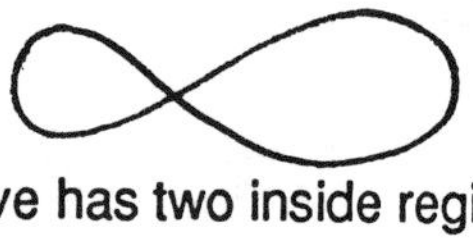

This curve has two inside regions.

This curve is not closed.
It also does not divide
the plane into one inside
and one outside region.

This situation gave rise to *the Jordan Theorem* – which states *that every closed curve in a plane which does not cross itself divides the plane into one inside region and one outside region.*

An easy method for determining whether a point is inside or outside a simple closed curve is illustrated by the example below. Present the problem to your students and have them discover how its location is linked to how many times the straight line crosses the curve.

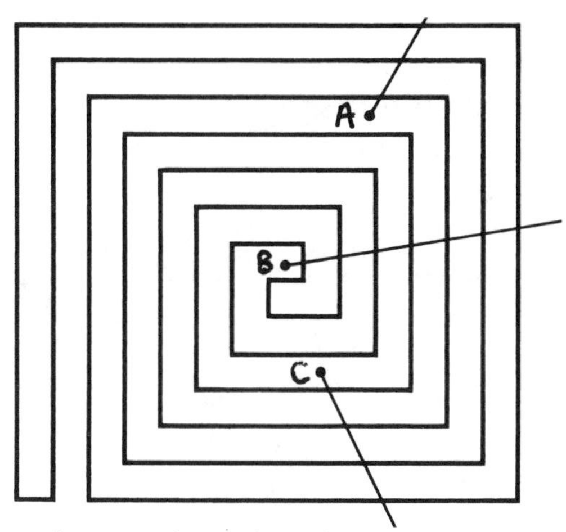

example: *To discover the method for determining the inside and outside regions for a point, draw a straight line from the point to the outside (in other words away from) the curve. Now count the number of times the straight line crosses the curve.*

If it crosses an even number of times the point is outside.
If it crosses an odd number of times the point is inside.

Write down the data with your students.

The line from point A crosses 3 times.
The line from point B crosses 7 times.
The line from point C crosses 4 times.

Then shade the inside region to visually distinguish it from the outside, and see if your students arrive at a connection between the number of crossings and the region where the point is located.

Test the conclusion on the other examples shown.

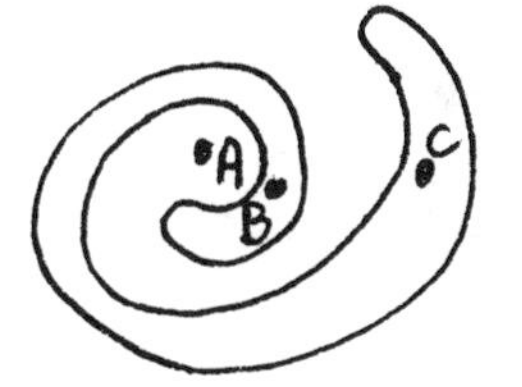

You may want to mention that the concept of simple closed curve can be extended to the 3rd dimension in space where it is called a simple closed surface. This surface now divides space into two regions–one inside and one outside.

examples: 1) this room
2) a sphere
3) the body's circulation system

Now discuss the topological concept of **transformation.** It has been a predominant theme throughout the lesson. In topology, if an object's classification does not change through distortions, then the changed object is called a transformation. For example, take a △ ,it can be distorted to a □ then to a ○ then to a , and through all the changes it remains a simple closed curve. It is the same topological figure. Transformations can change the size or shape of a figure, but they do not change the classifcation of the figure.

See if the students can distinguish which of the following changes are transformations.

examples:

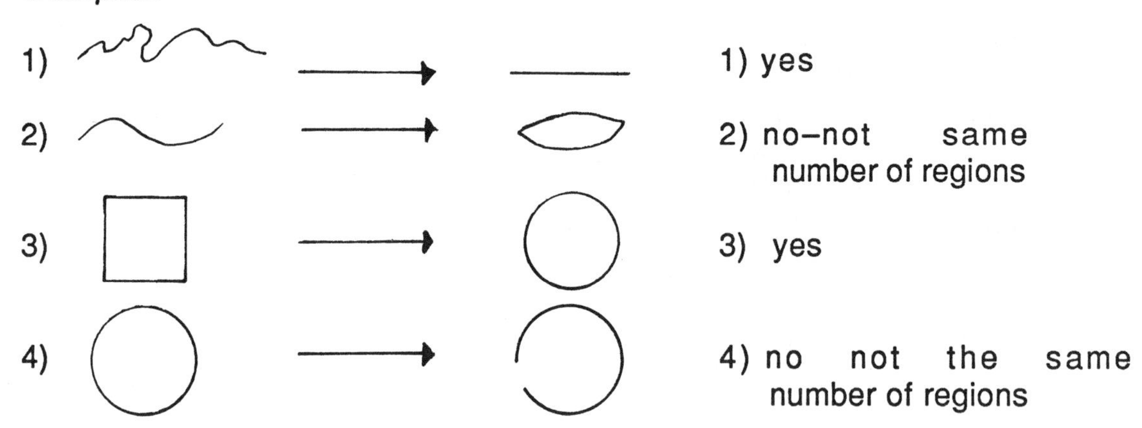

Other methods of classification–*optional*

1) For three dimensional objects, topology classifies objects by counting the number of cuts that would be needed to change it to a simple closed surface.

example: a doughnut shaped object would need one cut to change it into a sphere

2) An object can also be classified by the number of edges and surfaces it has.

examples: 1) a sheet of paper or a round card is considered to have one edge and two surfaces

2) an inner tube has no edge but two surfaces

3) Objects can also be classified by the number of cross-cuts (a cut that begins and ends on an edge) that can be made on a surface without dividing it into more than one piece.

Now that the students have some background on what topology is and what type of objects and properties it studies, it is appropriate to look at some of its fascinating problems.

Return to the problems presented at the beginning of the lesson, and consider their solutions with the students using topological terms and techniques.

(1) FLOOR PLAN PROBLEM: Derive a network with your students that can be used to solve the problem by the Euler technique of odd/even vertices.

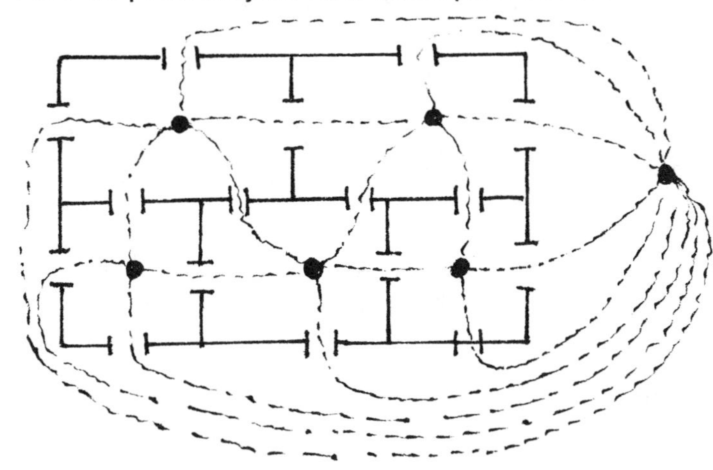

three 5-vertices
two 4-vertices
one 9-vertex

impossible because it has more than two odd vertices

(2) MOEBIUS STRIP: This fascinating topological model was created by the German mathematician Augustus Moebius (1790-1868). Suggested example and questions are:
When you look at a sheet of paper, how many sides does it have?

Do you think it is possible to distort this strip of paper, and make it end up having only one side?

Twist it and glue it together as shown.
Now we have a one-sided piece of paper.

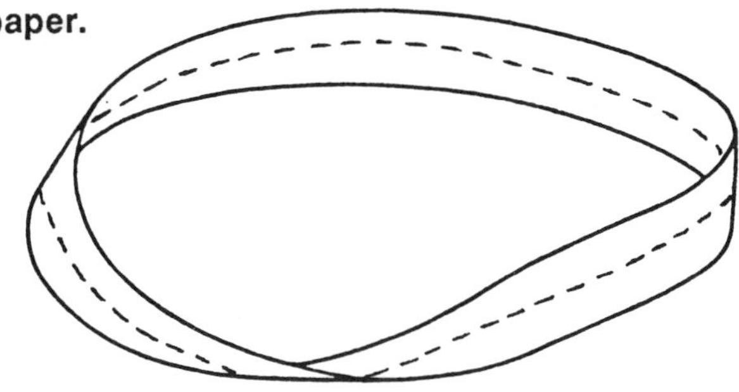

How do you prove it's only one-sided?
Take a crayon and draw a continous line until you return to the starting point without lifting your crayon.

Since this is possible, it has only one side. A second side would have necessitated your lifting the crayon.

The best way to illustrate **the properties of a Moebius strip** is to physically demonstrate the following for your students:

What happens to the strip of paper if instead of doing a 1/2 twist and gluing the ends together, you do two 1/2 twists? **Is it still one sided?**

Some students may ask you what practical use this knowledge has.
There is a tremendous amount of mathematics that exists that is not yet applicable to everyday life. Discoveries are being made continually that apply much of the mathematical knowledge that was once considered only abstract. The Moebius strip is one such bit of knowledge. If a belt, such as a fan belt on an automobile, were made in the shape of a Moebius strip, it would wear more evenly and last longer than the traditional fan belt because it would have only one side.

(3) THE CALIPH'S PROBLEM

A particular Persian Caliph devised a devious method for eliminating his daughter's potential suitors. He had two screening problems for the would be grooms.

He told the suitors to connect the number from one side to their respective number of the other side with a continuous curves that did not cross any other curve.

Nearly all the suitors solved the first problem, but none were able to solve the second one. Could the daughter ever hope to wed?

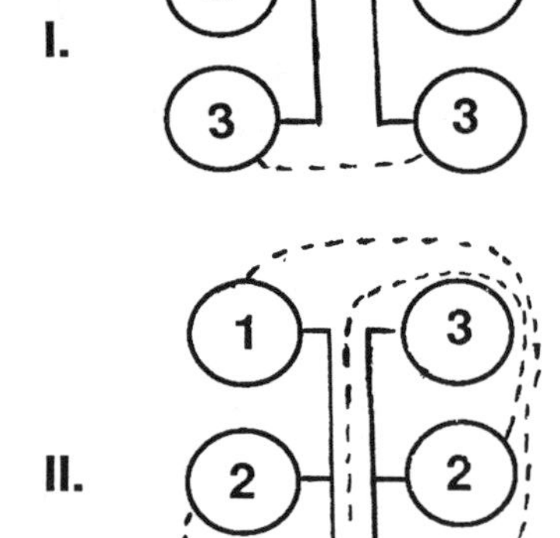

Discuss with your students why this problem involves the use of topological ideas. Show them that a simple closed curve is formed in problem (2) and that of the remaining numbers to be connected, one lies outside and the other inside this simple closed curve. So there is no way to join them on that plane without crossing a line.

(4) THE KLEIN BOTTLE

Now show the students an illustration of the Klein bottle on the overhead projector. The diagram shows how the Klein bottle is formed. It was invented by

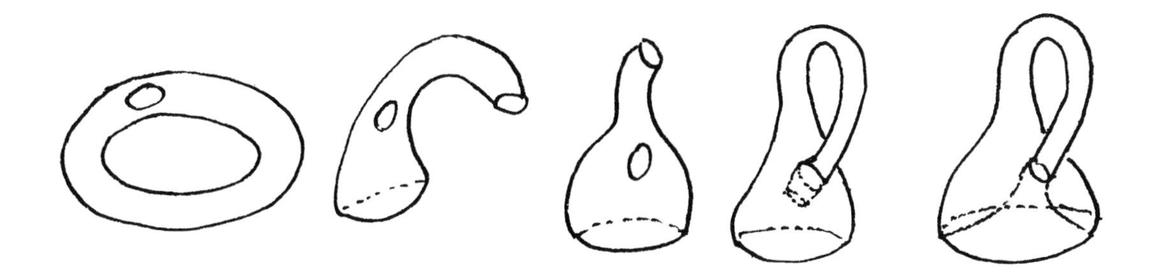

the German Mathematician, Felix Klein in 1882.

Point out that the Klein bottle is not a simple closed surface because it has only one surface or region. It has neither an inside, nor an outisde. (Note: The word inside can exist only if there is an outside and vice versa.)

Ask your class what would happen if water were put in the "hole" of this bottle. Now ask what would be formed if the Klein bottle were cut in half along its length. (Two Moebius strips would be formed, after twisting the two cut sections a bit.)

Summary

In discussing these four problems and objects we used topological ideas and concepts–

1) **one-sided and two sided objects on a *Moebius strip;***
2) **simple closed curve concept on the *Caliph problem;***
3) **inside, outside, simple closed surface, and distortion by cutting on the *Klein bottle.***

Review the concepts covered. Reemphasize that topology is a relatively new branch of mathematics which seeks to study the very basic similarities between objects. Review the definition of topology and the problems and ideas it studies. It is a young branch of mathematics that is expanding with the changing world.

LESSON'S OUTLINE

I. Introduction
  A. Network problem
  B. Koenigsberg Bridge Problem
    1) historical background
    2) Euler's solution using networks

II. Present various problems without solutions to stimulate curiosity
  A. the office problem
  B. Moebius strip
  C. Caliph problem
  D. Klein bottle

III. Topology
  A. definition
  B. equivalent figures
  C. elastic figures
    1) rubber sheet geometry

IV. Topological concepts
  A. Concepts topology deal with
    1) location
    2) inside vs. outside
    3) betweeness
  B. Concepts topology does not deal with
    1) length
    2) size
    3) shape
    4) distance

V. How topology categorizes objects
  A. Simple closed curve
    1) definition
    2) examples
    3) Jordan Theorem-optional
  B. Simple closed surface
    1) example
  C. Definition of transformation
    1) examples
  D. other classifications–optional
    1) cutting to form simple closed surface from 3-D object

2) number of edges and surfaces
3) cross-cuts

VII. Applications of topology and its concepts in discussing the problems and objects from II.

A. networks and the office problem
B. one-sided and two-sided objects and the Moebius strip
C. simple closed curve and the Caliph problem
D. the Klein bottle and inside, outside, simple closed surfaces and distortion

VIII. Summarize

A. What topology is and what it studies

IX. Pass out assignment

LESSON'S OBJECTIVES

1) Student will learn how the study of topology evolved.

2) Student will learn the definition of topology, simple closed curve, network, transformation, equivalent.

3) Student will learn the types of problems topology studies.

4) Student will learn what types of objects topology studies.

5) Student will learn some of the ways objects are classified in topology.

6) Student will learn how to apply some of the concepts of topology in problem solving.

ANSWERS to assignment on topology

1)
a) objects one, two, and four are equivalent
b) objects one, three, and four are equivalent
c) objects one, two ,and three are equivalent
d) objects, one, three, and four are equivalent
e) objects one and three are equivalent

2) one , three and four are traceable

3) (a) and (d)

4) A network diagram would show paths that would cross because a simple closed curve is formed. The only way to solve this is to go underground or above ground.

5) It is possbile because there are not more than two odd vertices, in fact all the vertices are even.

6) 1st–two cuts 2nd–no cuts 3rd–3 cuts

7) answers may vary by where bridge is added or deleted

9) 0, 4, 6, 9 are equivalent–they divide the region into two regions
1, 2, 3, 5, 7 are equivalent–they divide region into only one region
8–divides a region into three regions

10) The knot should disappear.

ASSIGNMENT

1) In each set, circle the figures that are topologically equivalent.

a)    

b)  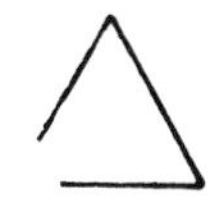  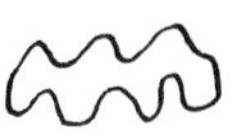

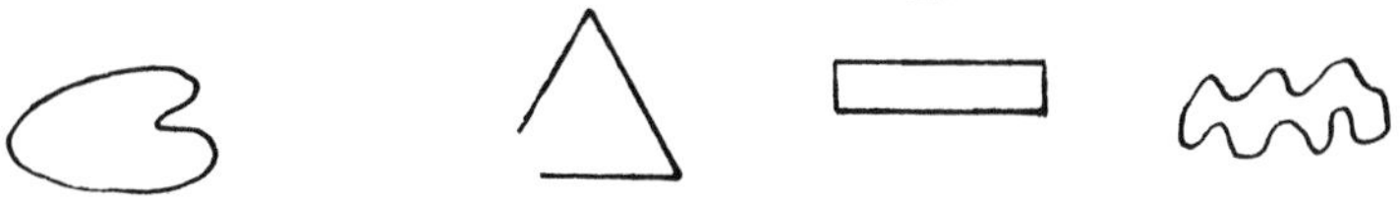

c) a sphere a balloon a box a circle

d)    

e) a glass a straw  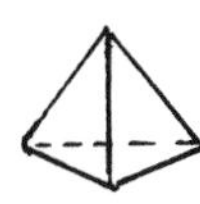

2) Which of the following networks are traceable?

 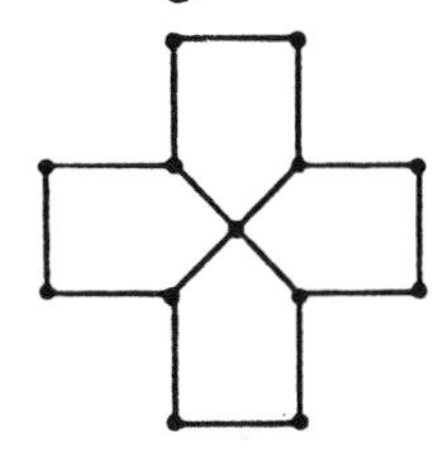  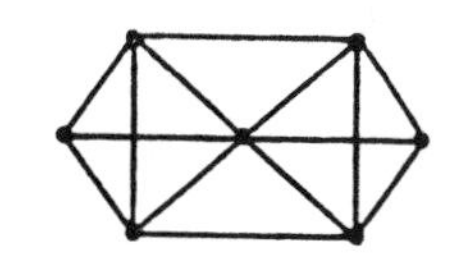 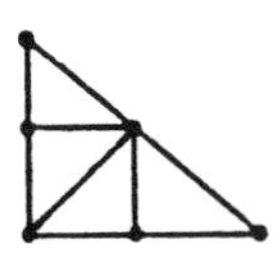

3) Which of the following properties do not change (are invariant) under a topological transformation?

a) number of insides and outsides
b) the length of a segment
c) the shape of an object
d) the arrangement of points on a line

4) From each home separate paths that lead to the well, to the grain mill, and to the woodshed must be made. Show why it is possible or impossible.

5) Draw a network of the following floor plan, and determine whether it is impossible to pass through each doorway one and only one time when walking through the house.

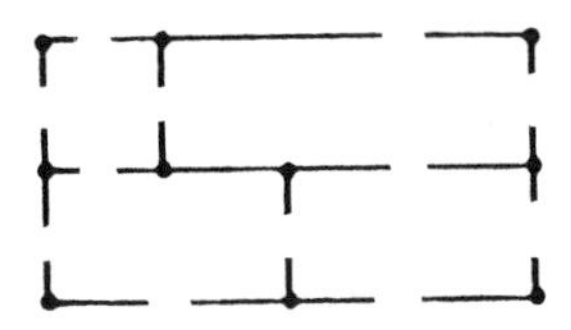

6) *Optional* –How many cuts are needed to change these 3-D objects to simple closed surfaces.

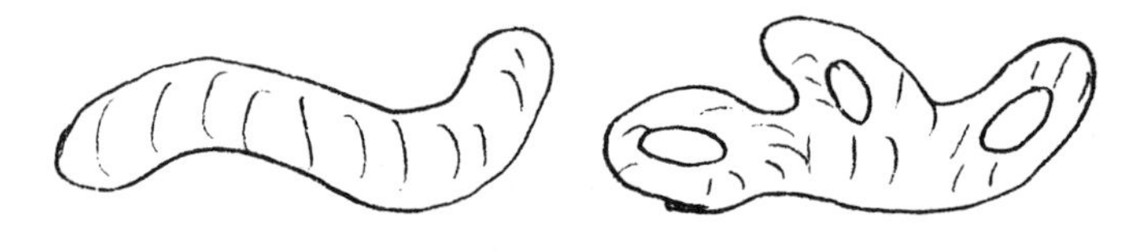

7) The diagram below is a picture of the Koenigsberg Bridge Problem. In class we showed how Euler used a network to prove it was not possible to find a path that would cross each bridge only once without doubling back. Change the diagram so that it is possible by removing or adding a bridge.

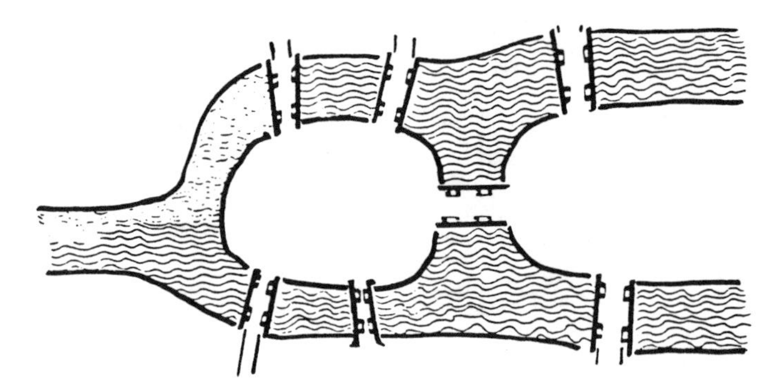

8) Verify your above answer by using a network to show whether it is possible or not to make a trip and cross over each bridge once.

9) Take the digits from 0 to 9, group them into topologically equivalent groups.

10) Take some string or rope and make the following knot. Pull both ends. What happens?

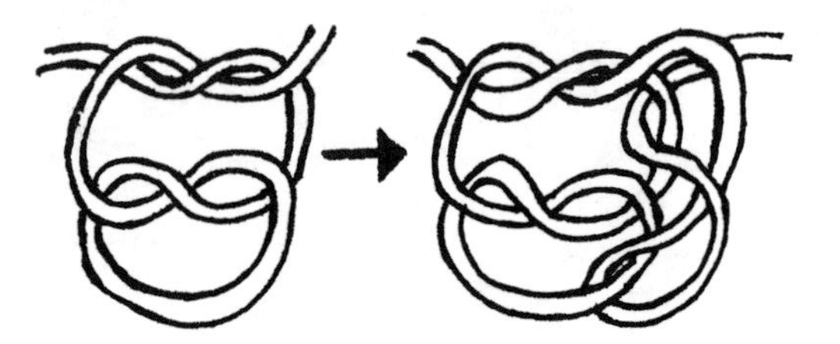

MAGIC SQUARES

52	61	4	13	20	29	36	45
14	3	62	51	46	35	30	19
53	60	5	12	21	28	37	44
11	6	59	54	43	38	27	22
55	58	7	10	23	26	39	42
9	8	57	56	41	40	25	24
50	63	2	15	18	31	34	47
16	1	64	49	48	33	32	17

MAGIC SQUARES

BACKGROUND MATERIAL

Magic squares have intrigued people for centuries. In fact, more has been written on magic squares than any other mathematical recreation. As their name implies, they were even connected to the occult.

The **order** of a magic square is defined by the number of rows or columns. For example, this magic square has order 3 because it has 3 rows.

8	1	6
3	5	7
4	9	2

The **magic** of a magic square arises from all the fascinating properties it possesses. Some of the **properties** are:

1) Each row, column, and diagonal total the same number. This magic constant can be obtained in one of the following ways:

a) Take the magic square's order, n, and find the value of $1/2[n(n^2+1)]$ where the magic square is composed of the natural numbers 1,2,3, . . ,n.

8	1	6
3	5	7
4	9	2

order 3
magic number=
$1/2(3(3^2+1))=15$

15 15

1	2	3
4	5	6
7	8	9

b) Take any size magic square. Starting from the lefthand corner, place the numbers sequentially along each row. The sum of the numbers in either diagonal will be the magic constant.

2) Any two numbers (in a row, column, or diagonal) that are equidistant from the center are complements. *Numbers of a magic square are complements if the sum is the same as the sum of the smallest and largest numbers of that magic square.*

8	1	6
3	5	7
4	9	2

complements
8 & 2
6 & 4
3 & 7
1 & 9

Ways to transform an existing magic square to another magic square:
3) Any number may be added to or mutiplied by every number of a magic square and it will remain a magic square.

4) If the two rows and two columns, equidistant from the center, are interchanged the resulting square is also a magic square.

5)
a) Interchanging quadrants in an even order magic square results in a magic square.
b) Interchanging partial quadrants and rows in an odd order magic square results in a magic square.

HISTORICAL BACKGROUND of magic squares

Magic squares were connected with the supernatural and the magical world from ancient times. Archeological excavations have turned them up in ancient Asian cities. In fact, the earliest record of the appearance of a magic square is *lo shu.* Legend has it that this magic square was first seen by Emperor Yu appearing on the back of a divine tortoise on the bank of the Yellow River about 2200 B.C. It was formed by knots in strings. The black knots are even numbers and the white ones are odd numers, and it forms a 3x3 magic square.

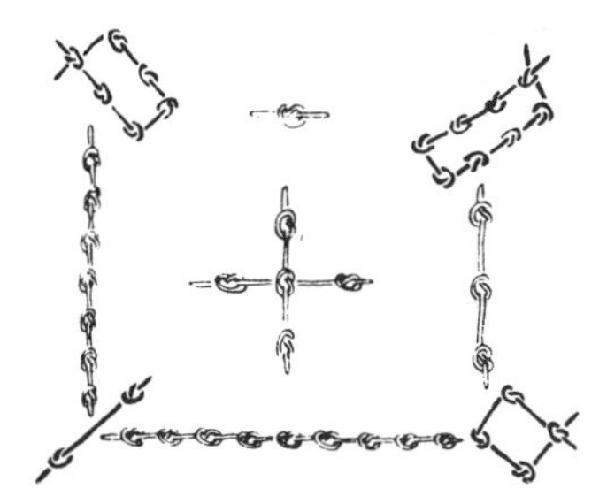

In the western world we find the first mention of magic squares in the work of Theon of Symrna in 130 A.D. In the 9th century, magic squares crept into the world of astrology with Arab astrologers using them in horoscope calculations. Finally, with the works of the Greek mathematician Moschopoulos, in 1300 A.D., magic squares and their respective properties spread to the western hemisphere (especially during the Renaissance period). To this day his work is in the National Library of Paris.

Magic squares have provided many with entertainment and challenge. A magic square even has a prominent place in a well known engraving, *Melencolia,* by Albrecht Durer in which he arranged the magic square to reveal the date of his work, 1514. Even Benjamin Franklin created his own magic square.

To study how magic squares are formed it is best to divide them into two classes–the odd and even order squares.

ODD ORDER MAGIC SQUARES

I. the staircase method

The staircase method can be used to construct odd order magic squares. This method was invented by La Loubère.

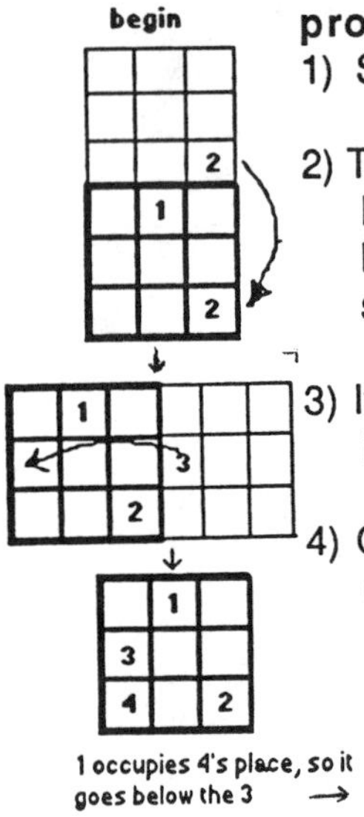

begin

procedure:

1) Start with the number 1 in the middle box of the top row.

2) The next number is placed diagonally upward in next box-unless it is occupied. If the box is in an imaginary magic square outside your magic square, find its location in your magic square by matching the location it holds in the imaginary square to your magic square.

3) If the diagonally upward box is occupied, then place the number in the box immediately below the original number.

4) Continue following steps (2) and (3) to obtain the location of the remaining numbers for the magic square.

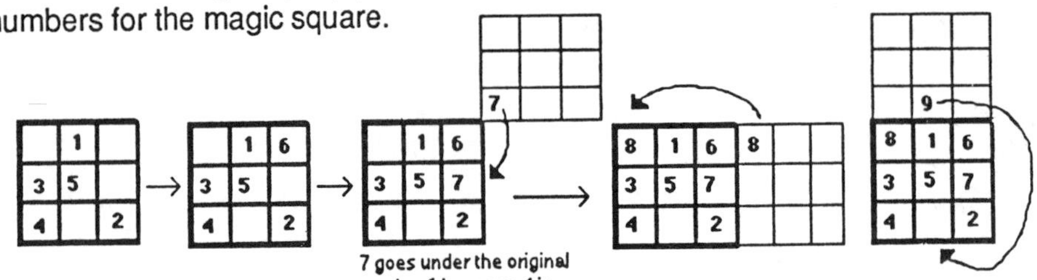

1 occupies 4's place, so it goes below the 3

7 goes under the original number 6 because 4 is occupying 7's place

II. The pyramid method is also used for constructing odd order magic squares. Actually this method should probably be called the diamond method since the numbers of the magic square are arranged in a diamond shape.

procedure:

1) Sequentially place the numbers of the magic square along diagonal boxes, as illustrated.
2) Relocate any numbers falling outside the magic square in imaginary squares, in their respective positions in the magic square.

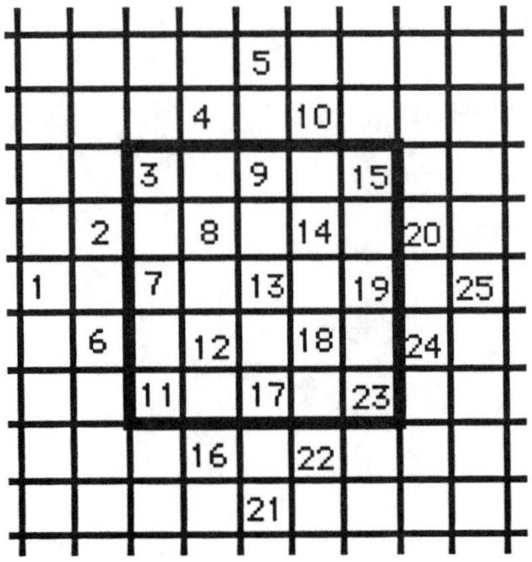

3	16	9	22	15
20	8	21	14	2
7	25	13	1	19
24	12	5	18	6
11	16	17	22	23

III. The knight's move method *(for odd orders of 5 or more)* forms an odd order magic square by selecting to place numbers in the same fashion as a knight may move in a chess game. For example, two places upward and one to the right.

procedure:

1) Place the number 1 in any box.

2) The next number is placed in a box two spaces upward and one to the right-if the box is vacant. If the box is outside the magic square in an imaginary square, relocate it in a matching position in the magic square.

3) If the box is not vacant place the number immediately below the original number.

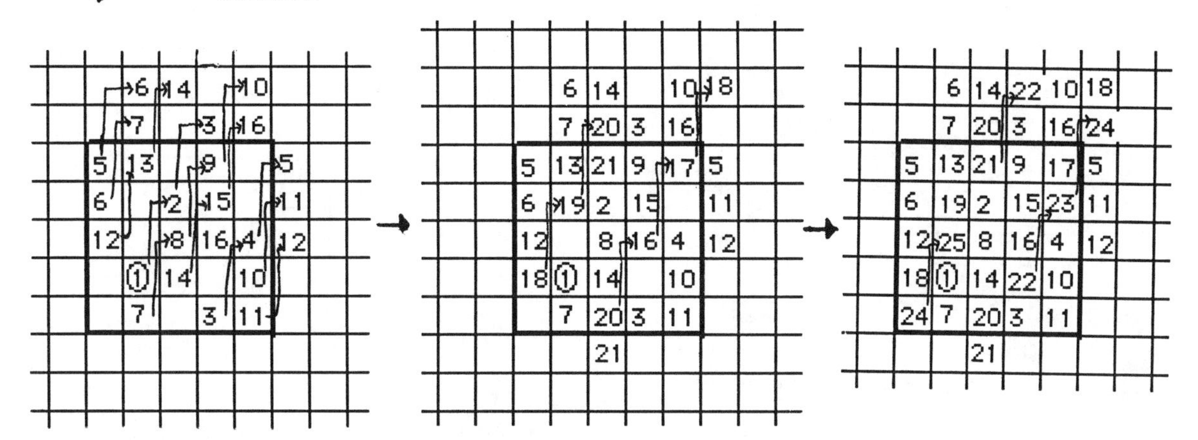

EVEN ORDER MAGIC SQUARES

Unlike the odd order, magic squares, where the methods discussed apply to any sized magic square, the even order magic squares do not have generalized methods of construction. They have to be individually developed for each even sized magic square.

example: the diagonal method applies only to a 4x4 magic square

procedure: Begin by sequentially placing the numbers in the rows of the magic square. If a number lands on one of the diagonals, its location must be switched with its complement.

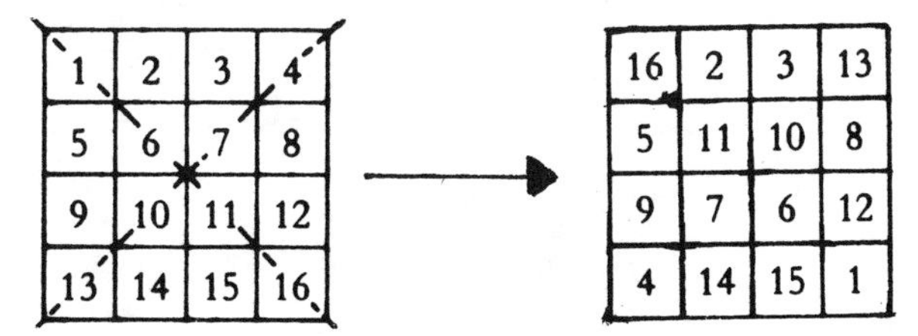

With a 4x4 magic square, either rows or columns may be interchanged, and it will remain a magic square. Also if quadrants are interchanged-it results in a magic square.

Constructing an 8x8 magic square

procedure: Sequentially place the 64 numbers in the rows as indicated. Draw in the indicated diagonals. Any numbers lying on diagonals should be interchanged with their complement.

1	2	3	4	5	6	7	8
9	10	11	12	13	14	15	16
17	18	19	20	21	22	23	24
25	26	27	28	29	30	31	32
33	34	35	36	37	38	39	40
41	42	43	44	45	46	47	48
49	50	51	52	53	54	55	56
57	58	59	60	61	62	63	64

→

64	2	3	61	60	6	7	57
9	55	54	12	13	51	50	16
17	47	46	20	21	43	42	24
40	26	27	37	36	30	31	33
32	34	35	29	28	38	39	25
41	23	22	44	45	19	18	48
49	15	14	52	53	11	10	56
8	58	59	5	4	62	63	1

Constructing a 12x12 magic square

procedure: a 12x12 magic square can be made by forming it into 16 composite 3x3 magic squares. There are 3 steps.

8	1	6
3	5	7
4	9	2

1) Form sixteen 3x3 magic squares–the first one using the first nine numerals, the second the next nine numerals, and so forth.
2) Now form a 4x4 magic square where its first box will correspond to the first 3x3 magic square, its second box to the second 3x3 magic square, and so on.

16	2	3	13
5	11	10	8
9	7	6	12
4	14	15	1

			17	10	15						
	16		12	14 **2**	16		**3**			**13**	
			13	18	11						
	5			**11**			**10**			**8**	
	9			**7**			**6**			**12**	
									8	1	6
	4			**14**			**15**		3	5 **1**	7
									4	9	2

3) Locate each 3x3 composite magic square into its respective box in the 4x4 magic square. The result is a 12x12 magic square.

143	136	141	17	10	15	26	19	24	116	109	114
138	140	142	12	14	16	21	23	25	111	113	115
139	144	137	13	18	11	22	27	20	112	117	110
44	37	42	98	91	96	89	82	87	71	64	69
39	41	43	93	95	97	84	86	88	66	68	70
40	45	38	94	99	92	85	90	83	67	72	65
80	73	78	62	55	60	53	46	51	107	100	105
75	77	79	57	59	61	48	50	52	102	104	106
76	81	74	58	63	56	49	54	47	103	108	101
35	28	33	125	118	123	134	127	132	8	1	6
30	32	34	120	122	124	129	131	133	3	5	7
31	36	29	121	126	119	130	135	128	4	9	2

As we see, constructing an even order magic square is very involved, since no general method exists. Various sizes have been constructed by magic square enthusiasts. A well known one is the 8x8 and 16x16 by Benjamin Franklin.

52	61	4	13	20	29	36	45
14	3	62	51	46	35	30	19
53	60	5	12	21	28	37	44
11	6	59	54	43	38	27	22
55	58	7	10	23	26	39	42
9	8	57	56	41	40	25	24
50	63	2	15	18	31	34	47
16	1	64	49	48	33	32	17

Benjamin Franklin's magic square features a variety of numerical oddities. Each row totals 260. Halfway totals 130. A shaded diagonal up four squares and down four squares totals 260. The sum of the four corners and the four center numbers is 260. The sum of the four numbers forming a little square (4x4) is 130. The sum of any four numbers equidistant from the center is 130.

Also related to the magic square is the ***magic line.*** Actually, a magic line is not a line, but is a pattern formed when consecutive numbers of a magic square are connected. Claude F. Bragdon (of the 1900's) discovered how magic squares could be used to form these artistically pleasing patterns with symmetry. As an architect, Bragdon used *magic lines* in architectural ornaments and graphic designs of books and textiles.

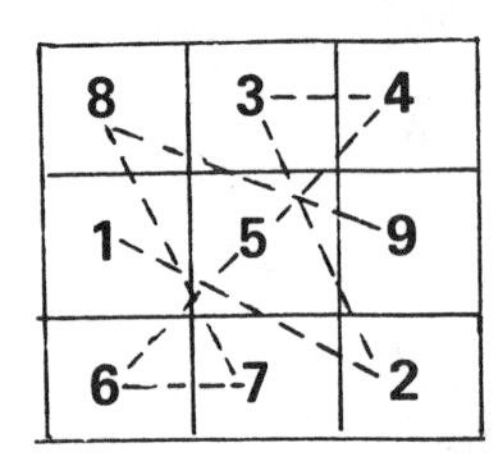

The magic line for lo shu, the earliest known magic square. It is from China about 2200 B.C.

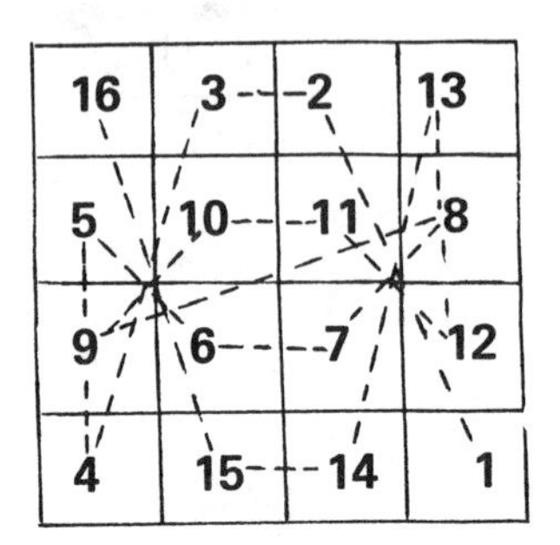

The magic line for Dürer's magic square of 1514.

Magic squares remain as fascinating and challenging today as they were to their discoverers of ancient times. The various ways of constructing them and their properties never cease to intrigue.

Melencolia by Albrecht Dürer, 1514

LESSON

An interesting way to introduce the lesson is to ask a student to give you a number (preferably a whole number), for example 22. With this number immediately construct a 3x3 magic square by using an odd order method mentioned in the background material, or by memorizing a 3x3 magic square. Try to make your presentation seem rather magical and mystical.

Once you have placed the magic square on the board or the overhead projector, have the students discover its magic number in all rows, columns and diagonals. Undoubtledy they will want to know how it is made. At this point you can lead into distinquishing between even order and odd order magic squares.

Using either a 3x3 or a 5x5 magic square, illustrate how to construct other magic squares by each of the methods discussed in the background material–staircase, pyramid, knight methods. To illustrate the various methods, make a graph grid transparency for the overhead projector using the grid provided. Now pass out 1/4 inch graph paper and ask them to make a magic square you have not made yet using one of the methods illustrated, and have them check for its magic number. You may want to show them how to calculate the magic number by one of the methods mentioned in the background material.

This is a good point to discuss the history of the magic square as covered in the background material–mentioning *lo-shu* as the earliest; showing the diagram on the overhead projector; the use of magic squares in horoscopes by Arab astrologers; how magic squares reached western Europe.

Once the students have successfuly made their magic square, discuss the properties of the magic square as detailed in the background material. Using a 3x3 or a 5x5 magic square, discover or illustrate these properties:

1) how to find the magic number for any magic square

2) any two numbers equidistant from the center will be complements

3) ways of altering a magic square into a new magic square:

   a) adding any number to each number of the magic square
   b) multiplying each number of the magic square by any selected number

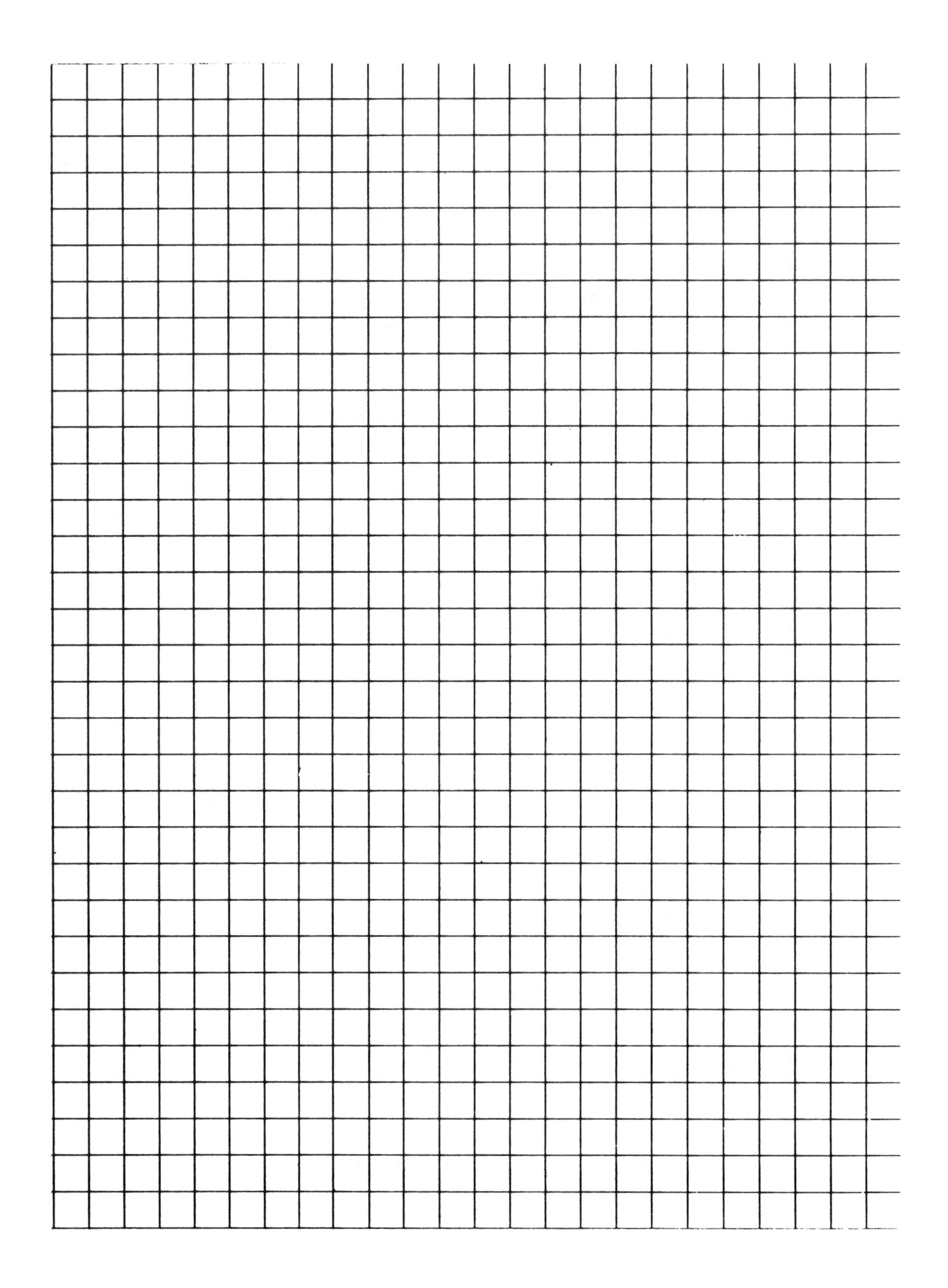

c) interchanging two columns and two rows equidistant from the center
d) interchanging quadrants in even order magic squares
e) interchanging partial quadrants and rows in odd order magic squares

example for (3a) adding any number to each number of the magic square and checking if the resulting square is a magic square

See if students discover a general rule, as the following for finding the new magic number. Since each number in the row is increased by ten the magic number should have increased by 30, here we have 15+30=45.

8	1	6
3	5	7
4	9	2

→

18	11	16
13	15	17
14	19	12

At this point you will need to assess your time,and decide if you want to discuss the other two methods for constructing odd order magic squares. It is important to discuss constructing even order magic squares, and to emphasize– *a general method has not been devised for constructing any size even order magic square as there has for the odd order ones.*

Refer to the background material for constructing the following even order magic squares.

1) 4x4 magic square by the diagonal method
2) 8x8 magic square by the multiple diagonals

Lastly, discuss the intriguing patterns that magic squares form called *magic lines* . Set up a 5x5 or a 4x4 or the ones illustrated in the background material on the board, and using a meter stick draw in their magic lines.

4	15	14	1
9	6	7	12
5	10	11	8
16	3	2	13

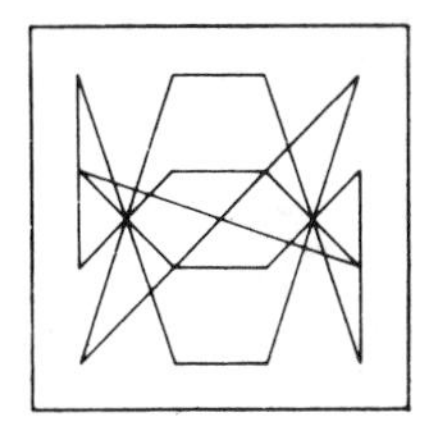

Finally, mention that there is a tremendous amount of material written on the recreation and mathematics of magic squares, and this lesson has only been an introduction to them.

LESSON'S OUTLINE

I. Introduction
- A. Construct a 3x3 magic square for your class from a random number given by a student.
- B magic number
- C. order of a magic square

II. Constructing odd order magic square
- A. use one of the general methods to construct either a 3x3 or a 5x5– staircase, pyramid, knight methods
  - 1) demonstrate the process on the overhead projector or on the board
  - 2) calculate its magic number by:
    - a) finding the total of a row, column, or diagonal
    - b) finding the value of $1/2n(n^2+1)$–*optional*
    - c) the diagonal method of finding the magic number

III. Historical background of magic square
- A. Asian cities
- B. lo-shu
- C. Theon of Smyrna
- D. Arab astrologers
- E. Greek mathematician, Moschopoulos–1300 A.D.
  - 1) introduced to western hemisphere
  - 2) work in National Library of Paris
  - 3) appearance in Renaissance

IV. Have students construct an odd order magic square on graph paper

V. Properties of a magic square
- A. illustrate by examples
  - 1) magic number–ways to determine
  - 2) complements
  - 3) equidistant numbers from center
  - 4) changing magic squares to new magic squares
    - a) adding or multiplying every number by a given number
    - b) interchanging two columns and two rows
    - c) interchanging columns and rows in specific ways for odd and even order magic squares

VI. Constructing even and odd magic squares
  A. no general method for even order magic square
    1) construct 4x4 using the diagonal method
    2) construct 8x8 using multiple diagonals
    3) discuss how a 12x12 would be constructed–*optional*

VII. magic lines
  A. construct a magic line for one of the magic squares
  B. developed by C. Bragdon
    1) patterns used by him in archirectural designs

VIII. Summary
  A. recreational and mathematical aspects of magic squares
  B. lesson is merely an introduction to the recreational and mathematical aspects of magic squares

IX. Pass out assignment.

LESSON'S OBJECTIVES

1) Student will learn the definition of a magic square.

2) Student will become familiar with the properties and terms of magic squares–magic number, order, even and odd order, complementary numbers, "magical" properties.

3) Student will learn how to construct any odd order magic square and a 4x4 magic square.

4) Student will learn the historical background of magic squares.

5) Student will learn how to change a magic square into a new magic square.

6) Student will experience the recreational and mathematical aspects of magic squares.

ASSIGNMENT

1) On quarter inch graph paper construct the following magic squares:

   a) using the numbers 1,2,3,. . .,25 for a 5x5 magic square

   b) using the number 1, 2,3,. . .,16 for a 4x4 magic square

2) a) Find the magic number for your 5x5 magic square in 1(a).________

   b) Find the magic number for your 4x4 magic square in 1(b).________

3) Pick a whole number_______. Now take your 5x5 magic square from 1(a) and add the number you picked to each number of your magic square.
   a) Write down the new magic square formed.

   b) What is this new magic square's magic number?_________

4) Pick a whole number________. Take your 4x4 magic square from 1(b) and multiply each of it numbers with the whole number you picked.

   a) Write down the new magic square formed.

   b) What is your new magic square's magic number?_________

5) a) Interchange two rows and write down a new magic square.

16	2	3	13
5	11	10	8
9	7	6	12
4	14	15	1

→ ?

b) Interchange partial quadrants A with D and B with C. Now interchange the row with the column that are forming a cross.
Is this new square a magic square?________ Verify your answer.

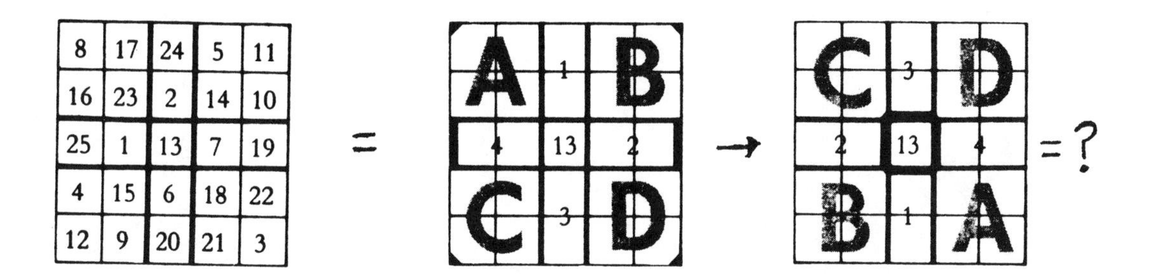

8	17	24	5	11
16	23	2	14	10
25	1	13	7	19
4	15	6	18	22
12	9	20	21	3

6) If you have a 9x9 magic square made up of the numbers 1,2,3,. . .,79,80,81, then what is its magic number?________

7) a) With a straightedge, draw the magic line for the 8x8 magic square below.

8	58	59	5	4	62	63	1
49	15	14	52	53	11	10	56
17	47	46	20	21	43	42	24
40	26	27	37	36	30	31	33
32	34	35	29	28	38	39	25
41	23	22	44	45	19	18	48
9	55	54	12	13	51	50	16
64	2	3	61	60	6	7	57

b) Divide the above 8x8 magic square into the following quadrants: Then switch the quadrants in the manner illustrated. The resulting square should be a magic square. **Draw its magic line.**

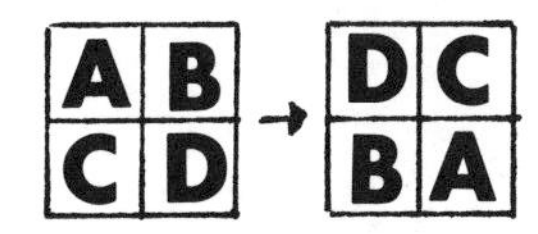

c) How do the magic lines in 7(a) and 7(b) compare in your opinion?

MATHEMATICAL LIBRARY ASSIGNMENT

MATHEMATICAL LIBRARY ASSIGNMENT

BACKGROUND MATERIAL

More often than not, libraries are surprised when a math class arrives to do research. Students need to learn how to learn and teach themselves, and where and how to gather information. Learning of available resources, how to use them, and methods of reseach are invaluable lessons and experiences. When an idea, term, or concept is unfamiliar, how and where does one find information on it? This lesson will provide the experience of researching ideas which are of an enrichment nature in mathematics. The concepts are rarely covered in detail, if at all, in traditional math courses which are focused on their specific curriculum.

The questions in the library assignment are varied, covering people, problems, ideas and concepts. The questions are also designed for short answers, but do not be surprised if you receive a long dissertation on one particular topic.

Students need to know that mathematics exists beyond the specific courses offered in the academic curricula–that indeed, mathematical knowledge, ideas, concepts and problems are so vast that mastering of coursework is not the world of mathematics. The library assignment isi intended to expose students to new ideas and to learn how to research them.

LESSON

Some advance preparation with students is needed before releasing them in the library. You will need to explain various resources in the library that can help in researching various questions in the library assignment. Some students may be experts on the use of the library, and others will not know what a periodical guide is or how to use the card catalogue. It is always helpful to arrange with the school librarian to give short presentation on using the library. The librarian may also have some prepared materials that can be given to the students before the assignment.

The assignment may be presented in different fashions. Adapt the most suitable for your class composition of students.

1) form research groups of 3 or 4 students
2) make it an individual assignment for each student
3) assign certain questions to certain students or groups, who will present their results to the class

I suggest you book the library for your class period. This will give the students time to get a good start on the assignment. In addition, you will be available for suggestions on research techniques.

There will probably be some questions students will not be successful in finding. Your school library may be limited, and you may want them to finish the assignment at the public library.

After the assignment is in, you may want to set aside class time to discuss various questions and answers.

The class discussion is a valuable reinforcement of the assignment and the concepts covered in the questions.

LIBRARY ASSIGNMENT

1) Who was Descartes, and what were some of his contributions to mathematics?

2) Why is the name Hilbert important to the study of geometry?

3) What is an imaginary number?

4) What does the letter **e** represent in mathematics?

5) What is the largest prime number discovered thus far?

6) Where did the numerals we use originate?

7) What are some of the beliefs of the ancient society of Pythagoreans?

8) What does the symbol π represent ?

9) a) How did the Babylonians write 8 and 25 ?
   b) How did the Romans write 8 and 25 ?
   c) How do the Chinese write 8 and 25 ?
   d) How did the Mayan write 8 and 25 ?

10) What is a googol and a googolplex?

11) What are "Napier's rods".

12) What does the word *geometry* mean?

13) What does, $\aleph_0$ read aleph null represent ?

14) What is Mersenne's number ?

15) What is projective geometry ?

16) Using only a compass and straightedge, what are the three ancient impossible construction problems of Euclidean geometry? Under what restrictions are they impossible?

17) a) What are at least two discoveries for which Archimedes is known?
   b) What were the unusual circumstances in his death?

18) What is a tesseract or a hypercube ?

19) Who was Hypatia, and how did she die ?

20) What is a quipu, and how was it used ?

LIBRARY ASSIGNMENT ANSWERS

These brief answers to the library assignment are intended to serve merely as a quick reference and do not represent the entire answer.

1) Who was Descartes, and what were some of his contributions to mathematics ?

René Descartes (1596-1650) was a mathematician and philosopher. In mathematics he is known for his invention of analytical geometry, which uses algebra to investigate and discover theorems of geometry. The Cartseian coordinate system is named after him.

He developed mathematical shorthand notation, and created Descartes' rule of signs, which set upper limit for the number of positive and negative roots for an equation. He also developed the follium of Descartes, which is a curve represented by the equation $x^3+y^3=3axy$.

2) Why is the name Hilbert important to the study of geometry?

David Hilbert (1862-1943) refined and took the bugs out of Euclid's geometry as set forth in *The Elements.*

3) What is an imaginary number?

i is defined as $\sqrt{-1}$. Any number with **i** in it is imaginary, e.g. 3i, -6i, 7+2i. **i** was invented when it was realized there were no real numbers that could be the square root of a negative number. **i** is the solution to the equation $x^2 = -1$.

4) What does the letter e represent in mathematics?

e represents a number that is approximately 2.71. . . It is an irrational transcendental number. It is the limit of the sequence, $\lim_{n \to \infty}\{1+1/n\}^n$ · It appears in many areas of mathematics. In banking, **e** can be used to calculate interest which is compounded continually.

5) What is the largest prime number discovered thus far?

As of May 1979, Harry Nelson of Liveromre Lab discovered $2^{4497}-1$. Yet a news report, on September 16, 1985, mentioned that a new prime number was computed on the Cray computer which is composed of 65,007 digits.(**Note:** verification of this report was not available at the time of this book's printing.)

6) Where did the numerals we use originate?

There are Hindu-Arabic numerals.

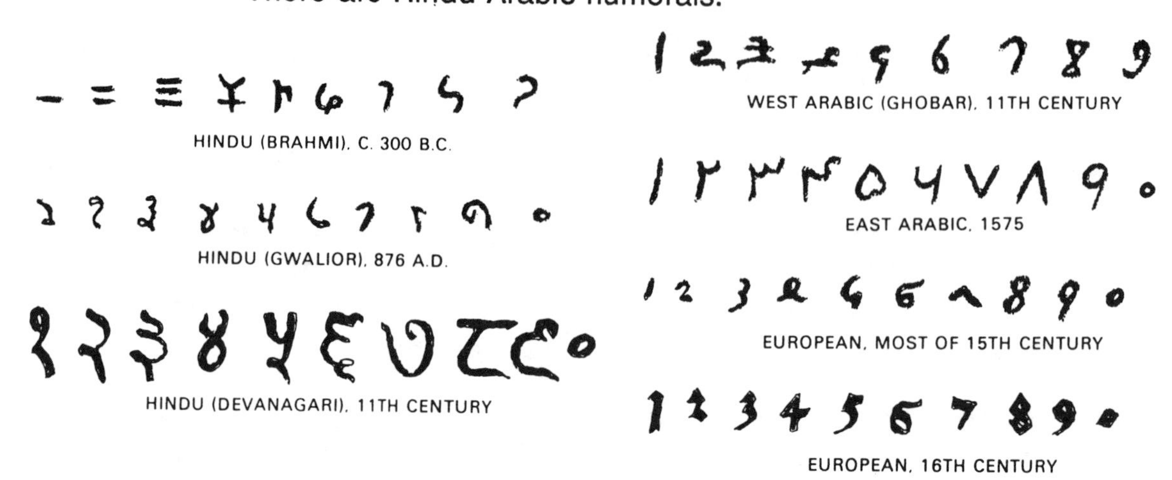

7) What are some of the beliefs of the ancient society of Pythagoreans?

They did not wear wool because they believed the souls of people were transfigured into animals, and that beans were sacred plants. Never take a higher road if a lower one is present (humility). Never poke a fire with iron because flame is the symbol of truth. *Numbers rule the universe.* Two numbers are called *amicable* if each is the sum of the divisors of the others, e.g. 284 and 220. A number is *perfect* if it is the sum of its proper divisors. A number is *deficient* if it exceeds the sum of its proper divisors. A number is *abundant* if it is less than the sum of its proper divisors. Pythagoras preached that everything depends on the whole numbers.

8) What does the symbol π represent in mathematics ?

π is defined as a circle's circumference divided by its diameter (π=C/d). From ancient times they had estimated its value around 3.14. Today we know it is an irrational transcendental number used in the computation of the circumference and area of any circle. Its value can now be approximated to any desired decimal place by the use of infinite series and computers.

9) a) How did the Babylonians write 8 and 25 ?________

b) How did the Romans write 8 and 25 ? VIII & XXV

c) How do the Chinese write 8 and 25 ?________

d) How did the Mayan write 8 and 25 ?________

10) What is a googol and a googolplex?

A googol is 1 followed by 100 zeros, 10^{100}. A googolplex is 1 followed by a googol of zeros, 10^{googol}. These numbers are used to described very large quantities, for example, if the universe were filled with protons and electrons so that no vacant space remained, the total number would be 10^{110}– more than a googol but less than a googolplex.

11) What are "Napier's bones".

In the 16th century, John Napier invented rods nicknamed "Napier's bones". Based on logarithms, they made calculations so easy that seamen could plot their course with little computation.

12) What does the word *geometry* mean?

Literally, to measure the earth.

13) What does $\aleph_0$, read aleph null represent ?

An ordinal number which represents the number of natural numbers.

14) What is Mersenne's number ?

It is the number–
132686104398972053177608575506095614293539359890333525802891469459697.

The 17th century sixty-nine digit number, proposed by French mathematician Marin Mersenne, was factored after 32 hours and 12 minutes of computer time. This feat now has cryptographers worried, since many cryptographic systems use multidigit numbers which are difficult to factor, to encode secrets and keep them secure. Its three factors are –

178230287214063289511 and 61676882198695257501367 and 12070396178249893039969681.

15) What is projective geometry ?

Projective geometry is a special kind of geometry that partly stemmed from the attempts of artists to add the illusion of depth to their paintings. This geometry is concerned with the properties of geometric figures that remain unchanged when they are projected onto another surface, e.g. another plane.

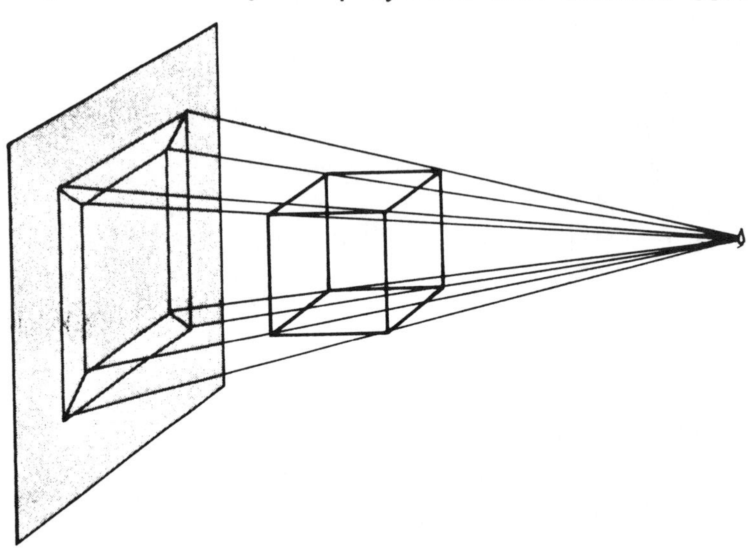

16) Using only a compass and straightedge, what are the three ancient impossible construction problems of Euclidean geometry? Under what restictions are they impossible?

trisecting an angle – dividing an angle into three congruent angles

duplicating a cube –constructing a cube with twice the volume of the given cube

squaring a circle –constructing a square with the same area as a given circle

These problems stimulated mathematical thought and discoveries for over 2000 years, until it was determined in the 19th century that the three construction problems were not possible using only a straightedge and compass.

With a straightedge and compass one can construct lines, segments, circles and arcs, which can be represented by linear or 2nd degree equations. The equations that arise from the three construction problems are either cubic or involve transcendental numbers, and thus cannot be done with a straightedge or compass.

17) a) What are at least two discoveries for which Archimedes is known?

Estimating the value of π ; devising a mathematical exponential system for writing very large numbers; *Archimedes' Principle* (a body immersed in fluid is bouyed up by a force equal to the weight of the displaced fluid); *Archimedes' screw* (a mechanical device used to direct water from a lower level -as a valley- to an upper level -as a mountain); and numerous other discoveries.

b) What were the unusual circumstances in his death?

There are numerous versions of how he was killed by a Roman soldier. One version has it that Archimedes was working on a mathematical problem when Syracuse was invaded by Marcus Claudius Marcellus. A Roman soldier entered his home and demanded that Archimedes follow him. Archimedes ignored him and continued to work on his problem. Angered, the soldier killed Archimedes with his sword.

18) What is a tesseract or a hypercube ?

It is a 4-dimensional representation of a cube. A cube drawn on paper is sketched in perspective to imply its 3-dimensional characteristic. Thus a tesseract drawn on paper is a perspective of a perspective.

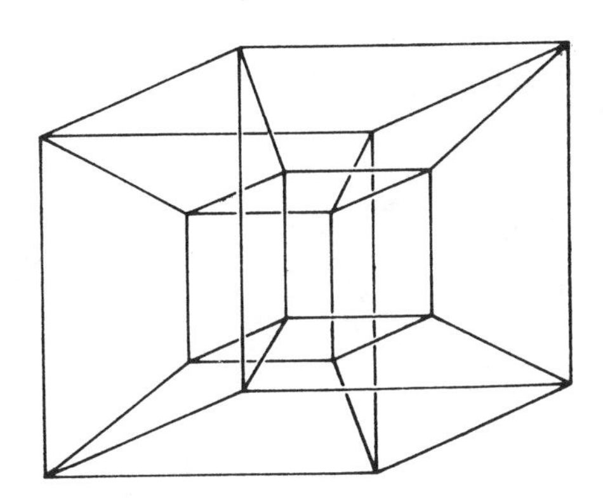

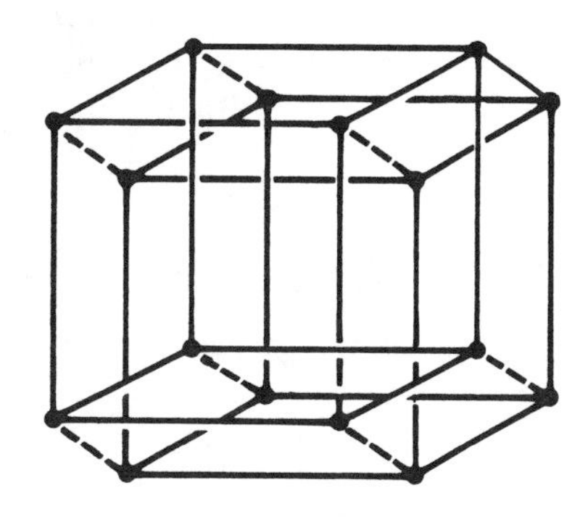

19) Who was Hypatia, and how did she die ?

Hypatia lived during the end of the 4th century A.D., a period of religious upheaval. She was the first woman mathematician to be mentioned in the history of mathematics. She studied and taught at Alexandria. She was well versed in mathematics, medicine and philosophy, and wrote commentaries on Diophantus' *Arithmetica* and Apollonius' *Conic Sections.* She lectured on Platonic philosophy at the Alexandrian Museum. Since she was a pagan, an angry mob of Christians siezed her and murdered her while she was at a pagan temple.

20) What is a quipu, and how was it used ?

The Inca empire encompassed the area around Cuzco, most of Peru, and parts of Ecuador and Chile. Although the Inca did not have a written mathematical notation or a written language, they managed their empire (more than two-thousand square miles long) by use of quipus. *Quipus* were knotted ropes using positional decimal system. A knot in the row farthest from the main strand represented one, next farthest represented 10, etc. No knots on the cord implied zero. The size, color and configuration of knots recorded information about crop yields, taxes, population, etc. Thus, these primitive computers–*quipus* – had knotted in their memory banks the information which tied together the Inca empire.

BIBLIOGRAPHY

Bakst, Aaron. Mathematics-***Its Magic and Mastery***. New York: D. Van Nostrand, Inc. 1952.

Bell, E.T. ***Mathematics Queen and Servant of Science.*** New York: McGraw-Hill Book Co., Inc., 1951.

Bergamini, David. ***Mathematics.*** New York: Time, Inc., 1963.

Bunt, Lucas; James, Phillip and Bedient. ***The Historical Roots of Elementary Mathematics.*** New Jersy: Prentice-Hall, Inc., 1976.

Campbell, Margaret W. ***Paper Toy Making.*** New York: Dover Publications, Inc., 1975.

Davis, Phililp J. and Hersh, Reuben. ***The Mathematical Experience.*** Boston: Houghton Mifflin Co., 1981.

Doczi, Gyorgy. ***The Power of Limits.*** Boulder, Colorado and London,England: Shambhala, 1981

Dudeney, H.E. ***Amusements in Mathematics.*** New York: Dover Publications, Inc.,1970.

Eves, Howard. ***An Introduction to the History of Mathematics.*** New York: Sanders College Publishing, 1983.

Freudenthal, Hans. ***Mathematics Observed.*** New York: Mcgraw-Hill Book Co., 1967.

Gardner, Martin. ***Aha !.*** New York: Scientific American, Inc./W.H. Freeman and Co., 1978.

Gardner, Martin. ***aha ! Gotcha.*** San Francisco: W.H. Freeman and Co., 1982.

Gardner, Martin. ***Mathematical Carnival.*** New York: Alfred A. Knopt, 1975.

Gardner, Martin. ***Mathematical Circus.*** New York: Alfred A. Knopt, 1979.

Gardner, Martin. ***New Mathematical Diversions.*** New York: Simon and Schuster, 1966.

Gardner, Martin. ***6th Book of Mathematical Games.*** New York: Charles Scribner's Sons, 1971.

Gardner, Martin. ***Mathematical Magic Show.*** New York: Alfred A. Knopt, 1977.

Gardner, Martin. ***Mathematical Puzzles and Diversions.*** New York: Simon and Schuster, 1961.

Ghyka, Matila. ***The Geometry of Art and Life.*** New York: Dover Publications, 1977.

Harblin, Robert. ***New Adventures in Origami.*** New York: Harper and Row, 1971.

Hogatt, Verner E. Jr. ***Fibonacci and Lucas Numbers.*** Boston: Houghton and Mifflin Co., 1969.

Hogben Lancelot. ***Mathematics in the Making.*** New York: Doubleday and Co., Inc., 1960.

Honda, Isao. ***The World of Origami.*** New York: Japan Publications Trading Co., 1965.

Johnson, Donovan A. and Glenn, William H. ***Exploring Mathematics on Your Own.*** New York: Doubleday and Co., Inc., 1961.

Kasner, Edward and Newman, James. ***Mathematics and The Imagination.*** New York: Simon and Schuster, 1940.

Kline, Morris. ***Mathematics Through from Ancient to Modern Times.*** New York: Oxford University Press, 1972.

Kline, Morris. Intro. ***Mathematics in the Modern World.*** San Francisco and London: W.H. Freeman and Co., 1968.

Kraitchik, Maurice. ***Mathematical Recreations.*** New York: Dover Publications, Inc., 1953.

Land, Frank. ***The Language of Mathematics.*** New York: Doubleday and Co., Inc., 1963.

Menninger, Karl. ***Number Words and Number Symbols.*** Cambridge, Massachusetts and London: The M.I.T. Press, 1969.

Meyer, Jerome S. ***Fun with Mathematics.*** New York: Fawcett World Library, 1957.

Moran, Jim. ***The Wonders of Magic Squares.*** New York: Random House, 1982.

Newman, James R. ed. ***The World of Mathematics.*** New York: Simon and Schuster, 1956.

Randlett, Samuel. ***The Best of Origami.*** New York: E.P. Dutton and Co., Inc.,1963.

Reid, Constance. ***From Zero to Infinity.*** New York: Thomas Y. Crowell Co., 1960.

Rogers, James T. ***The Story of Mathematics for Young People.*** New York: Random House, 1966.

Row, T. Sundara. ***Geometric Exercises in Paperfolding.*** New York: Dover Pub.,Inc.,1966.

Rucker, Rudy. ***Infinity and the Mind.*** Virginia: Randolph-Macon Woman's College, 1982.

Rucker, Rudolf. ***Geometry, Relativity and the Fourth Dimension.*** New York: Dover Publications, 1977.

Simon, William. ***Mathematical Magic.*** New York: Charles Scribner's Sons, 1964.

Sobel, Max A. and Maletsky, Evan M. ***Teaching Mathematics.***New Jersey: Prentice-Hall, Inc., 1975.

Steen, Lynn Arthur, ed. ***Mathematics Today.*** New York: Random House, 1980.
Steinhaus, H. ***Mathematical Snapshots.*** New York: Oxford University Press, 1969.
Stevens, Peters S. ***Patterns In Nature.*** Boston-Toronto: Little, Brown and Co.,1974.
Tuller, Anita. A ***Modern Introduction to Geometries.*** New Jersey: D.Van Nostrand Co.,Inc., 1968.
Zippin, Leo. ***Uses of Infinity.*** New York: Random House and the L.W. Singer Co., 1962.

ABOUT THE AUTHOR

Mathematics teacher and consultant Theoni Pappas received her B.A. from the University of California at Berkeley in 1966 and her M.A. from Stanford University in 1967.

Outside of teaching Ms. Pappas is best known for her unique and highly acclaimed **The Mathematics Calendar.** Her other creations include **The Math-T-Shirt, What Do you See?**–an optical illusion slide show with text, **The Children's Mathematics Calendar, The Mathematics Engagement Calendar, Greek Cooking for Everyone,** and **The Joy of Mathematics** – her most recent book – lets you discover some of the treasures of mathematics and the many places they exist.